INTER-GOVERNMENTAL MARITIME
CONSULTATIVE ORGANIZATION

INTERNATIONAL CONFERENCE

ON

MARINE POLLUTION, 1973

Final Act of the Conference, with
attachments, including

the

INTERNATIONAL CONVENTION FOR THE
PREVENTION OF POLLUTION FROM SHIPS, 1973

1977 Edition

LONDON

INTERNATIONAL CONFERENCE ON MARINE POLLUTION, 1973

TABLE OF CONTENTS

	Page
Final Act of the Conference	4
Attachment 1 — International Convention for the Prevention of Pollution from Ships, 1973	19
Protocol I — Provisions concerning Reports on Incidents involving Harmful Substances	31
Protocol II — Arbitration	33
Annex I — Regulations for the Prevention of Pollution by Oil	36
Appendix I — List of Oils	64
Appendix II — International Oil Pollution Prevention Certificate (1973)	65
Appendix III — Form of Oil Record Book	70
Annex II — Regulations for the Control of Pollution by Noxious Liquid Substances in Bulk	81
Appendix I — Guidelines for the Categorization of Noxious Liquid Substances	97
Appendix II — List of Noxious Substances carried in Bulk	98
Appendix III — List of other Liquid Substances carried in Bulk	104
Appendix IV — Cargo Record Book for Ships carrying Noxious Liquid Substances in Bulk	105
Appendix V — International Pollution Prevention Certificate for the Carriage of Noxious Liquid Substances in Bulk (1973)	108

INTERNATIONAL CONVENTION FOR THE PREVENTION OF POLLUTION FROM SHIPS, 1973

Procès-verbal of Rectification
13 June 1978

Rectifications have been made as follows:

ANNEX I

Page 43
Regulation 9(1)(a)(vi), lines 1/2 — insert "Regulation 15(5) and (6)" in place of "Regulation 15(3)"

Page 53
Regulation 16(6), line 4 — insert "less than" in place of "not more than"

Page 59
Regulation 24(1), line 1 — insert "provisions" in place of "provision"

Appendix III

II – FOR SHIPS OTHER THAN OIL TANKERS

Page 79
12(c) — insert "tank(s)" in place of "tank"

ANNEX II

Page 93
Regulation 9(3), line 1 — insert "Article 8" in place of "Article 7"

(Publication No. 77.14.E)

Page

Annex III — Regulations for the Prevention of Pollution by Harmful Substances carried by Sea in Packaged Forms, or in Freight Containers, Portable Tanks or Road and Rail Tank Wagons 111

Annex IV — Regulations for the Prevention of Pollution by Sewage from Ships 114

    Appendix — International Sewage Pollution Prevention Certificate (1973). 120

Annex V — Regulations for the Prevention of Pollution by Garbage from Ships 122

Attachment 3 — Resolutions 1-26 126

List of Persons attending the Conference 146

Conference Secretariat 168

NOTE

The text of Attachment 2 "Protocol Relating to Intervention on the High Seas in Cases of Marine Pollution by Substances other than Oil, 1973" is not reproduced in this edition, but has been included in another publication, together with the 1969 Convention Relating to Intervention on the High Seas in Cases of Oil Pollution Casualties (Sales number: 77.15.E).

FINAL ACT OF THE INTERNATIONAL CONFERENCE ON MARINE POLLUTION, 1973

1. By its Resolution A.176 (VI) of 21 October 1969, the Assembly of the Inter-Governmental Maritime Consultative Organization decided to convene in 1973 an International Conference on Marine Pollution. This Conference was held in London from 8 October to 2 November 1973.

2. The following States were represented by delegations at the Conference:

Argentina
Australia
Bahrain
Belgium
Brazil
Bulgaria
Byelorussian Soviet Socialist
    Republic
Canada
Chile
Cuba
Cyprus
Denmark
Dominican Republic
Ecuador
Egypt
Finland
France
German Democratic Republic
Germany, Federal Republic of
Ghana
Greece
Haiti
Hungary
Iceland
India
Indonesia
Iran
Iraq
Ireland
Italy
Ivory Coast
Japan
Jordan
Kenya
Khmer Republic
Kuwait
Liberia

Libyan Arab Republic
Madagascar
Mexico
Monaco
Morocco
Netherlands
New Zealand
Nigeria
Norway
Panama
Peru
Philippines
Poland
Portugal
Republic of Korea
Romania
Saudi Arabia
Singapore
South Africa
Spain
Sri Lanka
Sweden
Switzerland
Thailand
Trinidad and Tobago
Tunisia
Ukrainian Soviet Socialist
    Republic
Union of Soviet Socialist
    Republics
United Arab Emirates
United Kingdom of Great Britain
    and Northern Ireland
United Republic of Tanzania
United States of America
Uruguay
Venezuela

3. The following States were represented at the Conference by observers:

    Colombia   Republic of Viet-Nam
    Jamaica   Turkey
    Malawi   Yugoslavia
    Oman

    The Government of Hong Kong was also represented by an observer.

4. At the invitation of the Assembly the following organizations in the United Nations system sent representatives to the Conference:

    United Nations
    United Nations Environment Programme
    Food and Agriculture Organization
    United Nations Educational, Scientific and Cultural Organization
    International Bank for Reconstruction and Development
    International Atomic Energy Agency

5. The following inter-governmental organizations sent observers to the Conference:

    European Economic Community
    International Institute for the Unification of Private Law

6. The following non-governmental organizations also sent observers to the Conference:

    International Chamber of Shipping
    International Organization for Standardization
    International Electrotechnical Commission
    International Union of Marine Insurance
    International Association of Ports and Harbors
    The Baltic and International Maritime Conference
    International Association of Classification Societies
    International Law Association
    European Council of Chemical Manufacturers' Federation
    Oil Companies International Marine Forum
    International Shipowners' Association
    Friends of the Earth International

7. At the opening of the Conference The Hon. Michael Heseltine, Minister of Aerospace and Shipping of the United Kingdom and Mr. Maurice Strong, Executive Director of the United Nations Environment Programme made statements supporting the objectives of the Conference.

8. The Conference elected Mr. S. V. Bhave, Head of the Indian delegation, as President of the Conference.

9. Twenty-four Vice-Presidents of the Conference were elected, as follows:

    First Vice-President: Mr. G. Lindencrona (Sweden)

        Mr. R.M. Gowland (Argentina)
        H.E. Mr. M. Raffaelli (Brazil)

The Hon. Jack Davis (Canada)
Dr. M. Oporto (Cuba)
Mr. M.A. El-Sammak (Egypt)
Mr. J.P. Cabouat (France)
Dr. H. Rentner (German Democratic Republic)
Dr. G. Breuer (Germany, Federal Republic of)
H.E. Mr. H.V.H. Sekyi (Ghana)
Mr. M. Sjadzali (Indonesia)
Mr. H. Afshar (Iran)
Mr. K.G. Loukou (Ivory Coast)
H.E. Mr. S. Sugihara (Japan)
Mr. A.G. Toukan (Jordan)
Mr. E. Dinga (Kenya)
Mr. N.A. Al-Nakib (Kuwait)
Mr. M. Ramarozaka (Madagascar)
Dr. Vizcaíno Murray (Mexico)
Captain D.W. Boyes (New Zealand)
Mr. S. Perkowicz (Poland)
H.E. Mr. G. Nhigula (United Republic of Tanzania)
Mr. V. Tikhonov (USSR)
Mr. J.N. Archer (United Kingdom)

10. Mr. Colin Goad, Secretary-General of the Organization, acted as Secretary-General of the Conference with Mr. J. Quéguiner, Deputy Secretary-General, as Deputy Secretary-General of the Conference. Captain A. Saveliev, Secretary of the Maritime Safety Committee of the Organization, was appointed Executive Secretary of the Conference and Mr. Y. Sasamura, Head of Marine Science and Technology Division, and Mr. T. Mensah, Head of Legal Division, of the Organization were appointed Deputy Executive Secretaries of the Conference.

11. The Conference established the following Committees and a Steering Committee composed of officers of the Conference:

Committee I
    Chairman: H.E. Dr. P.V.J. Solomon (Trinidad and Tobago)
    Vice-Chairman: Mr. G. Lindencrona (Sweden)

Committee II
    Chairman: Dr. L. Spinelli (Italy)
    Vice-Chairman: Dr. W. Al-Nimer (Bahrain)

Committee III
    Chairman: Mr. R.J. Lakey (United States of America)
    Vice-Chairman: Mr. Koh Eng Tian (Singapore)

Committee IV
    Chairman: H.E. Prof. A. Yankov (Bulgaria)
    Vice-Chairman: The Hon. G.F.B. Cooper (Liberia)

Credentials Committee
    Chairman: Mr. P.A. Araque (Philippines)

Drafting Committee
    Chairman: Mr. G.A.E. Longe (Nigeria)
    Vice-Chairman: H.E. Mr. J.D. del Campo (Uruguay)

12. The following documentation formed the basis of the work of the Conference:

- Draft Text of an International Convention for the Prevention of Pollution from Ships, 1973

- Draft Protocol Relating to Intervention on the High Seas in Cases of Marine Pollution by Substances other than Oil

- Draft Resolutions relating to the prevention and control of marine pollution

- Proposals and comments, including amendments to the drafts mentioned above, submitted to the Conference by interested Governments and Organizations.

13. As a result of its deliberations, recorded in the summary records and reports of the Conference, the following instruments were adopted by the Conference:

INTERNATIONAL CONVENTION FOR THE PREVENTION OF POLLUTION FROM SHIPS, 1973

with its Protocols, Annexes and Appendices; and

PROTOCOL RELATING TO INTERVENTION ON THE HIGH SEAS IN CASES OF MARINE POLLUTION BY SUBSTANCES OTHER THAN OIL

The Convention and the Protocol constitute Attachments 1 and 2 to this Final Act respectively.*

14. The Conference also adopted Resolutions the texts of which comprise Attachment 3 of this Final Act.

15. The text of this Final Act including its attachments, is deposited with the Secretary-General of the Inter-Governmental Maritime Consultative Organization (IMCO). It is established in a single original in the English, French, Russian and Spanish languages, and accompanied by the texts of the International Convention for the Prevention of Pollution from Ships, 1973, with its Protocols, Annexes and Appendices, the Protocol relating to Intervention on the High Seas in Cases of Marine Pollution by Substances other than Oil, and the Resolutions of the Conference. The texts of the Convention, its Protocols, Annexes and Appendices, as well as of the Protocol, appear in their authentic languages, English, French, Russian and Spanish. The texts of Resolutions of the Conference appear in English, French, Russian and Spanish. Official translations of the Convention with its Protocols, Annexes and Appendices, and the Protocol, shall be prepared in the Arabic, German, Italian and Japanese languages. Originals of these official translations shall be deposited with this Final Act.

16. The Secretary-General of the Inter-Governmental Maritime Consultative Organization shall send a certified copy of this Final Act and, when they have been

---
* See NOTE on page 3.

prepared, certified copies of the official translations of the Convention with its Protocols, Annexes and Appendices, the Protocol and the Resolutions of the Conference to the Governments invited to be represented at the Conference in accordance with the wishes of those Governments.

IN WITNESS WHEREOF the undersigned have affixed their signatures to this Final Act.

DONE AT LONDON this second day of November, one thousand nine hundred and seventy-three.

President

S. V. BHAVE

Secretary-General of the Inter-Governmental Maritime Consultative Organization

COLIN GOAD

Deputy Secretary-General of the Inter-Governmental Maritime Consultative Organization

J. QUEGUINER

Executive Secretary of the Conference

A. SAVELIEV

Deputy Executive Secretary of the Conference

Y. SASAMURA

Deputy Executive Secretary of the Conference

THOMAS A. MENSAH

For the Government of the Argentine Republic

    R. M. GOWLAND

For the Government of the Commonwealth of Australia

    KEITH BRENNAN
    T. CURTIN

For the Government of the State of Bahrain

    W. NIMER

For the Government of the Kingdom of Belgium

    R. VANCRAEYNEST
    J. G. A. GERARD

For the Government of the Federative Republic of Brazil

    MARCELO RAFFAELLI

For the Government of the People's Republic of Bulgaria

    A. YANKOV

For the Government of the Byelorussian Soviet Socialist Republic

    V. PESHKOV

For the Government of Canada

    E. G. LEE
    R. W. PARSONS

For the Government of the Republic of Chile

    O. BUZETA

For the Government of the Republic of Cuba

    JUAN G. LOPEZ GARCIA
    M. OPORTO SANCHEZ

For the Government of the Republic of Cyprus

    MICHAEL V. VASSILIADES

For the Government of the Kingdom of Denmark

    G. SEIDENFADEN

For the Government of the Dominican Republic

    ALFREDO A. RICART

For the Government of the Republic of Ecuador

    J. ORTIZ

For the Government of the Arab Republic of Egypt

    MOSTAFA FAWZI

For the Government of the Republic of Finland

    E. HELANIEMI

For the Government of the French Republic

    J. MEGRET
    C. DOUAY

For the Government of the German Democratic Republic

    H. RENTNER

For the Government of the Federal Republic of Germany

    G. BREUER

For the Government of the Republic of Ghana

    H. V. H. SEKYI

For the Government of the Republic of Greece

    N. DIAMANTOPOULOS
    A. CHRONOPOULOS
    P. KOSMATOS

For the Government of the Republic of Haiti

For the Government of the Hungarian People's Republic

    G. KOTAI

For the Government of the Republic of Iceland

    HJALMAR R. BARDARSON
    HELGI AGUSTSSON

For the Government of the Republic of India

    S. V. BHAVE
    P. S. VANCHISWAR

For the Government of the Republic of Indonesia

    M. SJADZALI

For the Government of the Empire of Iran

For the Government of the Republic of Iraq

    A. AL-KADHI

For the Government of Ireland

    T. GORMAN

For the Government of the Italian Republic

    C. CALENDA

For the Government of the Republic of the Ivory Coast

For the Government of Japan

    S. SUGIHARA

For the Government of the Hashemite Kingdom of Jordan

    A. G. TOUKAN

For the Government of the Republic of Kenya

    R. O. ADERO

For the Government of the Khmer Republic

    SO YANDARA

For the Government of the State of Kuwait

    A. R. M. AL-YAGOUT

For the Government of the Republic of Liberia

    G. F. B. COOPER
    HENRY N. CONWAY Jr.
    F. L. WISWALL Jr.

For the Government of the Libyan Arab Republic

For the Government of the Malagasy Republic

    M. RAMAROZAKA

For the Government of the United Mexican States

    E. ECHEVERRIA

For the Government of the Principality of Monaco

    I. S. IVANOVIC

For the Government of the Kingdom of Morocco

For the Government of the Kingdom of the Netherlands

    J. F. VAN DOORN
    H. SONDAAL

For the Government of New Zealand

    DAVID W. BOYES

For the Government of the Federal Republic of Nigeria

    G. A. E. LONGE

For the Government of the Kingdom of Norway

    MODOLV HAREIDE
    E. HAREIDE

For the Government of the Republic of Panama

For the Government of the Republic of Peru

    A. MONTAGNE

For the Government of the Republic of the Philippines

    PABLO A. ARAQUE
    E. R. OGBINAR
    R. M. LESACA
    SONIA ZAIDE PRITCHARD

For the Government of the Polish People's Republic

    S. PERKOWICZ

For the Government of the Portuguese Republic

    M. C. CASQUINHO

For the Government of the Republic of Korea

    J. I. CHOI
    J. W. ROH

For the Government of the Socialist Republic of Romania

    V. STAN

For the Government of the Kingdom of Saudi Arabia

    FAISAL A. BASYONI

For the Government of the Republic of Singapore

    KOH ENG TIAN

For the Government of the Republic of South Africa

    J. J. BECKER

For the Government of the Spanish State

    ANTONIO POCH
    AMALIO GRAIÑO

For the Government of the Republic of Sri Lanka

    J. G. AMIRTHANAYAGAM

For the Government of the Kingdom of Sweden

    G. STEEN
    G. LINDENCRONA

For the Government of the Swiss Confederation

    R. BAR

For the Government of the Kingdom of Thailand

    K. SUPHAMONGKHON
    P. BURANADILOK
    P. CHARUCHANDR

For the Government of Trinidad and Tobago

    P. V. J. SOLOMON

For the Government of the Republic of Tunisia

    A. TURKI
    H. BOUSSOFFARA

For the Government of the Ukrainian Soviet Socialist Republic

    A. TRETIAK
    E. KACHURENKO

For the Government of the Union of Soviet Socialist Republics

    V. TIKHONOV

For the Government of the United Arab Emirates

For the Government of the United Kingdom of Great Britain and Northern Ireland

    J. N. ARCHER

For the Government of the United Republic of Tanzania

    S. IHEMA

For the Government of the United States of America

    R. E. TRAIN
    CHESTER R. BENDER
    W. M. BENKERT
    BERNARD H. OXMAN

For the Government of the Eastern Republic of Uruguay

    JUAN D. DEL CAMPO

For the Government of the Republic of Venezuela

    C. PEREZ DE LA COVA
    G. NOUT
    F. MARQUEZ
    R. HERNANDEZ

ATTACHMENT 1

INTERNATIONAL CONVENTION FOR THE PREVENTION OF POLLUTION FROM SHIPS, 1973

THE PARTIES TO THE CONVENTION,

BEING CONSCIOUS of the need to preserve the human environment in general and the marine environment in particular,

RECOGNIZING that deliberate, negligent or accidental release of oil and other harmful substances from ships constitutes a serious source of pollution,

RECOGNIZING ALSO the importance of the International Convention for the Prevention of Pollution of the Sea by Oil, 1954, as being the first multilateral instrument to be concluded with the prime objective of protecting the environment, and appreciating the significant contribution which that Convention has made in preserving the seas and coastal environment from pollution,

DESIRING to achieve the complete elimination of intentional pollution of the marine environment by oil and other harmful substances and the minimization of accidental discharge of such substances,

CONSIDERING that this object may best be achieved by establishing rules not limited to oil pollution having a universal purport,

HAVE AGREED as follows:

ARTICLE 1

General Obligations under the Convention

(1) The Parties to the Convention undertake to give effect to the provisions of the present Convention and those Annexes thereto by which they are bound, in order to prevent the pollution of the marine environment by the discharge of harmful substances or effluents containing such substances in contravention of the Convention.

(2) Unless expressly provided otherwise, a reference to the present Convention constitutes at the same time a reference to its Protocols and to the Annexes.

ARTICLE 2

Definitions

For the purposes of the present Convention, unless expressly provided otherwise:

(1) "Regulations" means the Regulations contained in the Annexes to the present Convention.

(2) "Harmful substance" means any substance which, if introduced into the sea, is liable to create hazards to human health, to harm living resources and marine life, to damage amenities or to interfere with other legitimate uses of the sea, and includes any substance subject to control by the present Convention.

(3) (a) "Discharge", in relation to harmful substances or effluents containing such substances, means any release howsoever caused from a ship and includes any escape, disposal, spilling, leaking, pumping, emitting or emptying;

    (b) "Discharge" does not include:

        (i) dumping within the meaning of the Convention on the Prevention of Marine Pollution by Dumping of Wastes and Other Matter, done at London on 13 November 1972; or

        (ii) release of harmful substances directly arising from the exploration, exploitation and associated off-shore processing of sea-bed mineral resources; or

        (iii) release of harmful substances for purposes of legitimate scientific research into pollution abatement or control.

(4) "Ship" means a vessel of any type whatsoever operating in the marine environment and includes hydrofoil boats, air-cushion vehicles, submersibles, floating craft and fixed or floating platforms.

(5) "Administration" means the Government of the State under whose authority the ship is operating. With respect to a ship entitled to fly a flag of any State, the Administration is the Government of that State. With respect to fixed or floating platforms engaged in exploration and exploitation of the sea-bed and subsoil thereof adjacent to the coast over which the coastal State exercises sovereign rights for the purposes of exploration and exploitation of their natural resources, the Administration is the Government of the coastal State concerned.

(6) "Incident" means an event involving the actual or probable discharge into the sea of a harmful substance, or effluents containing such a substance.

(7) "Organization" means the Inter-Governmental Maritime Consultative Organization.

ARTICLE 3

Application

(1) The present Convention shall apply to:

    (a) ships entitled to fly the flag of a Party to the Convention; and

    (b) ships not entitled to fly the flag of a Party but which operate under the authority of a Party.

(2) Nothing in the present Article shall be construed as derogating from or extending the sovereign rights of the Parties under international law over the sea-bed and subsoil thereof adjacent to their coasts for the purposes of exploration and exploitation of their natural resources.

(3) The present Convention shall not apply to any warship, naval auxiliary or other ship owned or operated by a State and used, for the time being, only on government non-commercial service. However, each Party shall ensure by the adoption of appropriate measures not impairing the operations or operational capabilities of such ships owned or operated by it, that such ships act in a manner consistent, so far as is reasonable and practicable, with the present Convention.

ARTICLE 4

Violation

(1) Any violation of the requirements of the present Convention shall be prohibited and sanctions shall be established therefor under the law of the Administration of the ship concerned wherever the violation occurs. If the Administration is informed of such a violation and is satisfied that sufficient evidence is available to enable proceedings to be brought in respect of the alleged violation, it shall cause such proceedings to be taken as soon as possible, in accordance with its law.

(2) Any violation of the requirements of the present Convention within the jurisdiction of any Party to the Convention shall be prohibited and sanctions shall be established therefor under the law of that Party. Whenever such a violation occurs, that Party shall either:

(a) cause proceedings to be taken in accordance with its law; or

(b) furnish to the Administration of the ship such information and evidence as may be in its possession that a violation has occurred.

(3) Where information or evidence with respect to any violation of the present Convention by a ship is furnished to the Administration of that ship, the Administration shall promptly inform the Party which has furnished the information or evidence, and the Organization, of the action taken.

(4) The penalties specified under the law of a Party pursuant to the present Article shall be adequate in severity to discourage violations of the present Convention and shall be equally severe irrespective of where the violations occur.

ARTICLE 5

Certificates and Special Rules on Inspection of Ships

(1) Subject to the provisions of paragraph (2) of the present Article a certificate issued under the authority of a Party to the Convention in accordance with the provisions of the Regulations shall be accepted by the other Parties and regarded for all purposes covered by the present Convention as having the same validity as a certificate issued by them.

(2) A ship required to hold a certificate in accordance with the provisions of the Regulations is subject, while in the ports or off-shore terminals under the jurisdiction of a Party, to inspection by officers duly authorized by that Party. Any such inspection shall be limited to verifying that there is on board a valid certificate, unless there are clear grounds for believing that the condition of the ship or its equipment

does not correspond substantially with the particulars of that certificate. In that case, or if the ship does not carry a valid certificate, the Party carrying out the inspection shall take such steps as will ensure that the ship shall not sail until it can proceed to sea without presenting an unreasonable threat of harm to the marine environment. That Party may, however, grant such a ship permission to leave the port or off-shore terminal for the purpose of proceeding to the nearest appropriate repair yard available.

(3) If a Party denies a foreign ship entry to the ports or off-shore terminals under its jurisdiction or takes any action against such a ship for the reason that the ship does not comply with the provisions of the present Convention, the Party shall immediately inform the consul or diplomatic representative of the Party whose flag the ship is entitled to fly, or if this is not possible, the Administration of the ship concerned. Before denying entry or taking such action the Party may request consultation with the Administration of the ship concerned. Information shall also be given to the Administration when a ship does not carry a valid certificate in accordance with the provisions of the Regulations.

(4) With respect to the ships of non-Parties to the Convention, Parties shall apply the requirements of the present Convention as may be necessary to ensure that no more favourable treatment is given to such ships.

ARTICLE 6

Detection of Violations and Enforcement of the Convention

(1) Parties to the Convention shall co-operate in the detection of violations and the enforcement of the provisions of the present Convention, using all appropriate and practicable measures of detection and environmental monitoring, adequate procedures for reporting and accumulation of evidence.

(2) A ship to which the present Convention applies may, in any port or off-shore terminal of a Party, be subject to inspection by officers appointed or authorized by that Party for the purpose of verifying whether the ship has discharged any harmful substances in violation of the provisions of the Regulations. If an inspection indicates a violation of the Convention, a report shall be forwarded to the Administration for any appropriate action.

(3) Any Party shall furnish to the Administration evidence, if any, that the ship has discharged harmful substances or effluents containing such substances in violation of the provisions of the Regulations. If it is practicable to do so, the competent authority of the former Party shall notify the Master of the ship of the alleged violation.

(4) Upon receiving such evidence, the Administration so informed shall investigate the matter, and may request the other Party to furnish further or better evidence of the alleged contravention. If the Administration is satisfied that sufficient evidence is available to enable proceedings to be brought in respect of the alleged violation, it shall cause such proceedings to be taken in accordance with its law as soon as possible. The Administration shall promptly inform the Party which has reported the alleged violation, as well as the Organization, of the action taken.

(5) A Party may also inspect a ship to which the present Convention applies when it enters the ports or off-shore terminals under its jurisdiction, if a request for an

investigation is received from any Party together with sufficient evidence that the ship has discharged harmful substances or effluents containing such substances in any place. The report of such investigation shall be sent to the Party requesting it and to the Administration so that the appropriate action may be taken under the present Convention.

ARTICLE 7

Undue Delay to Ships

(1) All possible efforts shall be made to avoid a ship being unduly detained or delayed under Articles 4, 5 or 6 of the present Convention.

(2) When a ship is unduly detained or delayed under Article 4, 5 or 6 of the present Convention, it shall be entitled to compensation for any loss or damage suffered.

ARTICLE 8

Reports on Incidents Involving Harmful Substances

(1) A report of an incident shall be made without delay to the fullest extent possible in accordance with the provisions of Protocol I to the present Convention.

(2) Each Party to the Convention shall:

(a) make all arrangements necessary for an appropriate officer or agency to receive and process all reports on incidents; and

(b) notify the Organization with complete details of such arrangements for circulation to other Parties and Member States of the Organization.

(3) Whenever a Party receives a report under the provisions of the present Article, that Party shall relay the report without delay to:

(a) the Administration of the ship involved; and

(b) any other State which may be affected.

(4) Each Party to the Convention undertakes to issue instructions to its maritime inspection vessels and aircraft and to other appropriate services, to report to its authorities any incident referred to in Protocol I to the present Convention. That Party shall, if it considers it appropriate, report accordingly to the Organization and to any other party concerned.

ARTICLE 9

Other Treaties and Interpretation

(1) Upon its entry into force, the present Convention supersedes the International Convention for the Prevention of Pollution of the Sea by Oil, 1954, as amended, as between Parties to that Convention.

(2) Nothing in the present Convention shall prejudice the codification and development of the law of the sea by the United Nations Conference on the Law of the Sea convened pursuant to Resolution 2750 C(XXV) of the General Assembly of the United Nations nor the present or future claims and legal views of any State concerning the law of the sea and the nature and extent of coastal and flag State jurisdiction.

(3) The term "jurisdiction" in the present Convention shall be construed in the light of international law in force at the time of application or interpretation of the present Convention.

ARTICLE 10

Settlement of Disputes

Any dispute between two or more Parties to the Convention concerning the interpretation or application of the present Convention shall, if settlement by negotiation between the Parties involved has not been possible, and if these Parties do not otherwise agree, be submitted upon request of any of them to arbitration as set out in Protocol II to the present Convention.

ARTICLE 11

Communication of Information

(1) The Parties to the Convention undertake to communicate to the Organization:

(a) the text of laws, orders, decrees and regulations and other instruments which have been promulgated on the various matters within the scope of the present Convention;

(b) a list of non-governmental agencies which are authorized to act on their behalf in matters relating to the design, construction and equipment of ships carrying harmful substances in accordance with the provisions of the Regulations;

(c) a sufficient number of specimens of their certificates issued under the provisions of the Regulations;

(d) a list of reception facilities including their location, capacity and available facilities and other characteristics;

(e) official reports or summaries of official reports in so far as they show the results of the application of the present Convention; and

(f) an annual statistical report, in a form standardized by the Organization, of penalties actually imposed for infringement of the present Convention.

(2) The Organization shall notify Parties of the receipt of any communications under the present Article and circulate to all Parties any information communicated to it under sub-paragraphs (1)(b) to (f) of the present Article.

ARTICLE 12

Casualties to Ships

(1) Each Administration undertakes to conduct an investigation of any casualty occurring to any of its ships subject to the provisions of the Regulations if such casualty has produced a major deleterious effect upon the marine environment.

(2) Each Party to the Convention undertakes to supply the Organization with information concerning the findings of such investigation, when it judges that such information may assist in determining what changes in the present Convention might be desirable.

ARTICLE 13

Signature, Ratification, Acceptance, Approval and Accession

(1) The present Convention shall remain open for signature at the Headquarters of the Organization from 15 January 1974 until 31 December 1974 and shall thereafter remain open for accession. States may become Parties to the present Convention by:

- (a) signature without reservation as to ratification, acceptance or approval; or
- (b) signature subject to ratification, acceptance or approval, followed by ratification, acceptance or approval; or
- (c) accession.

(2) Ratification, acceptance, approval or accession shall be effected by the deposit of an instrument to that effect with the Secretary-General of the Organization.

(3) The Secretary-General of the Organization shall inform all States which have signed the present Convention or acceded to it of any signature or of the deposit of any new instrument of ratification, acceptance, approval or accession and the date of its deposit.

ARTICLE 14

Optional Annexes

(1) A State may at the time of signing, ratifying, accepting, approving or acceding to the present Convention declare that it does not accept any one or all of Annexes III, IV and V (hereinafter referred to as "Optional Annexes") of the present Convention. Subject to the above, Parties to the Convention shall be bound by any Annex in its entirety.

(2) A State which has declared that it is not bound by an Optional Annex may at any time accept such Annex by depositing with the Organization an instrument of the kind referred to in Article 13(2).

(3) A State which makes a declaration under paragraph (1) of the present Article in respect of an Optional Annex and which has not subsequently accepted that Annex in accordance with paragraph (2) of the present Article shall not be under any obligation nor entitled to claim any privileges under the present Convention in

respect of matters related to such Annex and all references to Parties in the present Convention shall not include that State in so far as matters related to such Annex are concerned.

(4) The Organization shall inform the States which have signed or acceded to the present Convention of any declaration under the present Article as well as the receipt of any instrument deposited in accordance with the provisions of paragraph (2) of the present Article.

ARTICLE 15

Entry into Force

(1) The present Convention shall enter into force twelve months after the date on which not less than 15 States, the combined merchant fleets of which constitute not less than fifty per cent of the gross tonnage of the world's merchant shipping, have become parties to it in accordance with Article 13.

(2) An Optional Annex shall enter into force twelve months after the date on which the conditions stipulated in paragraph (1) of the present Article have been satisfied in relation to that Annex.

(3) The Organization shall inform the States which have signed the present Convention or acceded to it of the date on which it enters into force and of the date on which an Optional Annex enters into force in accordance with paragraph (2) of the present Article.

(4) For States which have deposited an instrument of ratification, acceptance, approval or accession in respect of the present Convention or any Optional Annex after the requirements for entry into force thereof have been met but prior to the date of entry into force, the ratification, acceptance, approval or accession shall take effect on the date of entry into force of the Convention or such Annex or three months after the date of deposit of the instrument whichever is the later date.

(5) For States which have deposited an instrument of ratification, acceptance, approval or accession after the date on which the Convention or an Optional Annex entered into force, the Convention or the Optional Annex shall become effective three months after the date of deposit of the instrument.

(6) After the date on which all the conditions required under Article 16 to bring an amendment to the present Convention or an Optional Annex into force have been fulfilled, any instrument of ratification, acceptance, approval or accession deposited shall apply to the Convention or Annex as amended.

ARTICLE 16

Amendments

(1) The present Convention may be amended by any of the procedures specified in the following paragraphs.

(2) Amendments after consideration by the Organization:

(a) any amendment proposed by a Party to the Convention shall be submitted to the Organization and circulated by its Secretary-General to all Members of the Organization and all Parties at least six months prior to its consideration;

(b) any amendment proposed and circulated as above shall be submitted to an appropriate body by the Organization for consideration;

(c) Parties to the Convention, whether or not Members of the Organization, shall be entitled to participate in the proceedings of the appropriate body;

(d) amendments shall be adopted by a two-thirds majority of only the Parties to the Convention present and voting;

(e) if adopted in accordance with sub-paragraph (d) above, amendments shall be communicated by the Secretary-General of the Organization to all the Parties to the Convention for acceptance;

(f) an amendment shall be deemed to have been accepted in the following circumstances:

(i) an amendment to an Article of the Convention shall be deemed to have been accepted on the date on which it is accepted by two-thirds of the Parties, the combined merchant fleets of which constitute not less than fifty per cent of the gross tonnage of the world's merchant fleet;

(ii) an amendment to an Annex to the Convention shall be deemed to have been accepted in accordance with the procedure specified in sub-paragraph (f)(iii) unless the appropriate body, at the time of its adoption, determines that the amendment shall be deemed to have been accepted on the date on which it is accepted by two-thirds of the Parties, the combined merchant fleets of which constitute not less than fifty per cent of the gross tonnage of the world's merchant fleet. Nevertheless, at any time before the entry into force of an amendment to an Annex to the Convention, a Party may notify the Secretary-General of the Organization that its express approval will be necessary before the amendment enters into force for it. The latter shall bring such notification and the date of its receipt to the notice of Parties;

(iii) an amendment to an Appendix to an Annex to the Convention shall be deemed to have been accepted at the end of a period to be determined by the appropriate body at the time of its adoption, which period shall be not less than ten months, unless within that period an objection is communicated to the Organization by not less than one-third of the Parties or by the Parties the combined merchant fleets of which constitute not less than fifty per cent of the gross tonnage of the world's merchant fleet whichever condition is fulfilled;

(iv) an amendment to Protocol I to the Convention shall be subject to the same procedures as for the amendments to the Annexes to the Convention, as provided for in sub-paragraphs (f)(ii) or (f)(iii) above;

- (v) an amendment to Protocol II to the Convention shall be subject to the same procedures as for the amendments to an Article of the Convention, as provided for in sub-paragraph (f)(i) above;

(g) the amendment shall enter into force under the following conditions:

- (i) in the case of an amendment to an Article of the Convention, to Protocol II, or to Protocol I or to an Annex to the Convention not under the procedure specified in sub-paragraph (f)(iii), the amendment accepted in conformity with the foregoing provisions shall enter into force six months after the date of its acceptance with respect to the Parties which have declared that they have accepted it;
- (ii) in the case of an amendment to Protocol I, to an Appendix to an Annex or to an Annex to the Convention under the procedure specified in sub-paragraph (f)(iii), the amendment deemed to have been accepted in accordance with the foregoing conditions shall enter into force six months after its acceptance for all the Parties with the exception of those which, before that date, have made a declaration that they do not accept it or a declaration under sub-paragraph (f)(ii), that their express approval is necessary.

(3) Amendment by a Conference:

(a) Upon the request of a Party, concurred in by at least one-third of the Parties, the Organization shall convene a Conference of Parties to the Convention to consider amendments to the present Convention.

(b) Every amendment adopted by such a Conference by a two-thirds majority of those present and voting of the Parties shall be communicated by the Secretary-General of the Organization to all Contracting Parties for their acceptance.

(c) Unless the Conference decides otherwise, the amendment shall be deemed to have been accepted and to have entered into force in accordance with the procedures specified for that purpose in paragraph (2)(f) and (g) above.

(4) (a) In the case of an amendment to an Optional Annex, a reference in the present Article to a "Party to the Convention" shall be deemed to mean a reference to a Party bound by that Annex.

(b) Any Party which has declined to accept an amendment to an Annex shall be treated as a non-Party only for the purpose of application of that Amendment.

(5) The adoption and entry into force of a new Annex shall be subject to the same procedures as for the adoption and entry into force of an amendment to an Article of the Convention.

(6) Unless expressly provided otherwise, any amendment to the present Convention made under this Article, which relates to the structure of a ship, shall apply only to ships for which the building contract is placed, or in the absence of a building contract, the keel of which is laid, on or after the date on which the amendment comes into force.

(7) Any amendment to a Protocol or to an Annex shall relate to the substance of that Protocol or Annex and shall be consistent with the Articles of the present Convention.

(8) The Secretary-General of the Organization shall inform all Parties of any amendments which enter into force under the present Article, together with the date on which each such amendment enters into force.

(9) Any declaration of acceptance or of objection to an amendment under the present Article shall be notified in writing to the Secretary-General of the Organization. The latter shall bring such notification and the date of its receipt to the notice of the Parties to the Convention.

ARTICLE 17

Promotion of Technical Co-operation

The Parties to the Convention shall promote, in consultation with the Organization and other international bodies, with assistance and co-ordination by the Executive Director of the United Nations Environment Programme, support for those Parties which request technical assistance for:

(a) the training of scientific and technical personnel;

(b) the supply of necessary equipment and facilities for reception and monitoring;

(c) the facilitation of other measures and arrangements to prevent or mitigate pollution of the marine environment by ships; and

(d) the encouragement of research;

preferably within the countries concerned, so furthering the aims and purposes of the present Convention.

ARTICLE 18

Denunciation

(1) The present Convention or any Optional Annex may be denounced by any Parties to the Convention at any time after the expiry of five years from the date on which the Convention or such Annex enters into force for that Party.

(2) Denunciation shall be effected by notification in writing to the Secretary-General of the Organization who shall inform all the other Parties of any such notification received and of the date of its receipt as well as the date on which such denunciation takes effect.

(3) A denunciation shall take effect twelve months after receipt of the notification of denunciation by the Secretary-General of the Organization or after the expiry of any other longer period which may be indicated in the notification.

ARTICLE 19

Deposit and Registration

(1) The present Convention shall be deposited with the Secretary-General of the Organization who shall transmit certified true copies thereof to all States which have signed the present Convention or acceded to it.

(2) As soon as the present Convention enters into force, the text shall be transmitted by the Secretary-General of the Organization to the Secretary-General of the United Nations for registration and publication, in accordance with Article 102 of the Charter of the United Nations.

ARTICLE 20

Languages

The present Convention is established in a single copy in the English, French, Russian and Spanish languages, each text being equally authentic. Official translations in the Arabic, German, Italian and Japanese languages shall be prepared and deposited with the signed original.

IN WITNESS WHEREOF the undersigned* being duly authorized by their respective Governments for that purpose have signed the present Convention.

DONE AT LONDON this second day of November, one thousand nine hundred and seventy-three.

---

* *Signatures omitted.*

PROTOCOL I

**PROVISIONS CONCERNING REPORTS ON
INCIDENTS INVOLVING HARMFUL SUBSTANCES**
(in accordance with Article 8 of the Convention)

Article I

Duty to Report

(1) The Master of a ship involved in an incident referred to in Article III of this Protocol, or other person having charge of the ship, shall report the particulars of such incident without delay and to the fullest extent possible in accordance with the provisions of this Protocol.

(2) In the event of the ship referred to in paragraph (1) of the present Article being abandoned, or in the event of a report from such ship being incomplete or unobtainable, the owner, charterer, manager or operator of the ship, or their agents shall, to the fullest extent possible assume the obligations placed upon the Master under the provisions of this Protocol.

Article II

Methods of Reporting

(1) Each report shall be made by radio whenever possible, but in any case by the fastest channels available at the time the report is made. Reports made by radio shall be given the highest possible priority.

(2) Reports shall be directed to the appropriate officer or agency specified in paragraph (2)(a) of Article 8 of the Convention.

Article III

When to make Reports

The report shall be made whenever an incident involves:

(a) a discharge other than as permitted under the present Convention; or

(b) a discharge permitted under the present Convention by virtue of the fact that:

　　(i) it is for the purpose of securing the safety of a ship or saving life at sea; or

　　(ii) it results from damage to the ship or its equipment; or

(c) a discharge of a harmful substance for the purpose of combating a specific pollution incident or for purposes of legitimate scientific research into pollution abatement or control; or

(d) the probability of a discharge referred to in sub-paragraphs (a), (b) or (c) of this Article.

Article IV

Contents of Report

(1) Each report shall contain in general:

    (a) the identity of the ship;

    (b) the time and date of the occurrence of the incident;

    (c) the geographic position of the ship when the incident occurred;

    (d) the wind and sea conditions prevailing at the time of the incident; and

    (e) relevant details respecting the condition of the ship.

(2) Each report shall contain, in particular:

    (a) a clear indication or description of the harmful substances involved, including, if possible, the correct technical names of such substances (trade names should not be used in place of the correct technical names);

    (b) a statement or estimate of the quantities, concentrations and likely conditions of harmful substances discharged or likely to be discharged into the sea;

    (c) where relevant, a description of the packaging and identifying marks; and

    (d) if possible the name of the consignor, consignee or manufacturer.

(3) Each report shall clearly indicate whether the harmful substance discharged, or likely to be discharged is oil, a noxious liquid substance, a noxious solid substance or a noxious gaseous substance and whether such substance was or is carried in bulk or contained in packaged form, freight containers, portable tanks, or road and rail tank wagons.

(4) Each report shall be supplemented as necessary by any other relevant information requested by a recipient of the report or which the person sending the report deems appropriate.

Article V

Supplementary Report

Any person who is obliged under the provisions of this Protocol to send a report shall, when possible:

    (a) supplement the initial report, as necessary, with information concerning further developments; and

    (b) comply as fully as possible with requests from affected States for additional information concerning the incident.

PROTOCOL II

ARBITRATION
(in accordance with Article 10 of the Convention)

Article I

Arbitration procedure, unless the Parties to the dispute decide otherwise, shall be in accordance with the rules set out in this Protocol.

Article II

(1) An Arbitration Tribunal shall be established upon the request of one Party to the Convention addressed to another in application of Article 10 of the present Convention. The request for arbitration shall consist of a statement of the case together with any supporting documents.

(2) The requesting Party shall inform the Secretary-General of the Organization of the fact that it has applied for the establishment of a Tribunal, of the names of the Parties to the dispute, and of the Articles of the Convention or Regulations over which there is in its opinion disagreement concerning their interpretation or application. The Secretary-General shall transmit this information to all Parties.

Article III

The Tribunal shall consist of three members: one Arbitrator nominated by each Party to the dispute and a third Arbitrator who shall be nominated by agreement between the two first named, and shall act as its Chairman.

Article IV

(1) If, at the end of a period of sixty days from the nomination of the second Arbitrator, the Chairman of the Tribunal shall not have been nominated, the Secretary-General of the Organization upon request of either Party shall within a further period of sixty days proceed to such nomination, selecting him from a list of qualified persons previously drawn up by the Council of the Organization.

(2) If, within a period of sixty days from the date of the receipt of the request, one of the Parties shall not have nominated the member of the Tribunal for whose designation it is responsible, the other Party may directly inform the Secretary-General of the Organization who shall nominate the Chairman of the Tribunal within a period of sixty days, selecting him from the list prescribed in paragraph (1) of the present Article.

(3) The Chairman of the Tribunal shall, upon nomination, request the Party which has not provided an Arbitrator, to do so in the same manner and under the same conditions. If the Party does not make the required nomination, the Chairman of the Tribunal shall request the Secretary-General of the Organization to make the nomination in the form and conditions prescribed in the preceding paragraph.

(4) The Chairman of the Tribunal, if nominated under the provisions of the present Article, shall not be or have been a national of one of the Parties concerned, except with the consent of the other Party.

(5) In the case of the decease or default of an Arbitrator for whose nomination one of the Parties is responsible, the said Party shall nominate a replacement within a period of sixty days from the date of decease or default. Should the said Party not make the nomination, the arbitration shall proceed under the remaining Arbitrators. In case of the decease or default of the Chairman of the Tribunal, a replacement shall be nominated in accordance with the provisions of Article III above, or in the absence of agreement between the members of the Tribunal within a period of sixty days of the decease or default, according to the provisions of the present Article.

Article V

The Tribunal may hear and determine counter-claims arising directly out of the subject matter of the dispute.

Article VI

Each Party shall be responsible for the remuneration of its Arbitrator and connected costs and for the costs entailed by the preparation of its own case. The remuneration of the Chairman of the Tribunal and of all general expenses incurred by the Arbitration shall be borne equally by the Parties. The Tribunal shall keep a record of all its expenses and shall furnish a final statement thereof.

Article VII

Any Party to the Convention which has an interest of a legal nature and which may be affected by the decision in the case may, after giving written notice to the Parties which have originally initiated the procedure, join in the arbitration procedure with the consent of the Tribunal.

Article VIII

Any Arbitration Tribunal established under the provisions of the present Protocol shall decide its own rules of procedure.

Article IX

(1) Decisions of the Tribunal both as to its procedure and its place of meeting and as to any question laid before it, shall be taken by majority votes of its members; the absence or abstention of one of the members of the Tribunal for whose nomination the Parties were responsible, shall not constitute an impediment to the Tribunal reaching a decision. In cases of equal voting, the vote of the Chairman shall be decisive.

(2) The Parties shall facilitate the work of the Tribunal and in particular, in accordance with their legislation, and using all means at their disposal:

    (a) provide the Tribunal with the necessary documents and information;

    (b) enable the Tribunal to enter their territory, to hear witnesses or experts, and to visit the scene.

(3) Absence or default of one Party shall not constitute an impediment to the procedure.

Article X

(1) The Tribunal shall render its award within a period of five months from the time it is established unless it decides, in the case of necessity, to extend the time limit for a further period not exceeding three months. The award of the Tribunal shall be accompanied by a statement of reasons. It shall be final and without appeal and shall be communicated to the Secretary-General of the Organization. The Parties shall immediately comply with the award.

(2) Any controversy which may arise between the Parties as regards interpretation or execution of the award may be submitted by either Party for judgment to the Tribunal which made the award, or, if it is not available to another Tribunal constituted for this purpose, in the same manner as the original Tribunal.

ANNEX I

REGULATIONS FOR THE PREVENTION OF POLLUTION BY OIL

CHAPTER I — GENERAL

Regulation 1

Definitions

For the purposes of this Annex:

(1) "Oil" means petroleum in any form including crude oil, fuel oil, sludge, oil refuse and refined products (other than petrochemicals which are subject to the provisions of Annex II of the present Convention) and, without limiting the generality of the foregoing, includes the substances listed in Appendix I to this Annex.

(2) "Oily mixture" means a mixture with any oil content.

(3) "Oil fuel" means any oil used as fuel in connexion with the propulsion and auxiliary machinery of the ship in which such oil is carried.

(4) "Oil tanker" means a ship constructed or adapted primarily to carry oil in bulk in its cargo spaces and includes combination carriers and any "chemical tanker" as defined in Annex II of the present Convention when it is carrying a cargo or part cargo of oil in bulk.

(5) "Combination carrier" means a ship designed to carry either oil or solid cargoes in bulk.

(6) "New ship" means a ship:

    (a) for which the building contract is placed after 31 December 1975; or

    (b) in the absence of a building contract, the keel of which is laid or which is at a similar stage of construction after 30 June 1976; or

    (c) the delivery of which is after 31 December 1979; or

    (d) which has undergone a major conversion:

        (i) for which the contract is placed after 31 December 1975; or

        (ii) in the absence of a contract, the construction work of which is begun after 30 June 1976; or

        (iii) which is completed after 31 December 1979.

(7) "Existing ship" means a ship which is not a new ship.

(8) "Major conversion" means a conversion of an existing ship:

(a) which substantially alters the dimensions or carrying capacity of the ship; or

(b) which changes the type of the ship; or

(c) the intent of which in the opinion of the Administration is substantially to prolong its life; or

(d) which otherwise so alters the ship that if it were a new ship, it would become subject to relevant provisions of the present Convention not applicable to it as an existing ship.

(9) "Nearest land". The term "from the nearest land" means from the baseline from which the territorial sea of the territory in question is established in accordance with international law, except that, for the purposes of the present Convention "from the nearest land" off the north eastern coast of Australia shall mean from a line drawn from a point on the coast of Australia in

latitude 11°00' South, longitude 142°08' East to a point in latitude 10°35' South,
longitude 141°55' East — thence to a point latitude 10°00' South,
longitude 142°00' East, thence to a point latitude 9°10' South,
longitude 143°52' East, thence to a point latitude 9°00' South,
longitude 144°30' East, thence to a point latitude 13°00' South,
longitude 144°00' East, thence to a point latitude 15°00' South,
longitude 146°00' East, thence to a point latitude 18°00' South,
longitude 147°00' East, thence to a point latitude 21°00' South,
longitude 153°00' East, thence to a point on the coast of Australia in latitude 24°42' South, longitude 153°15' East.

(10) "Special area" means a sea area where for recognized technical reasons in relation to its oceanographical and ecological condition and to the particular character of its traffic the adoption of special mandatory methods for the prevention of sea pollution by oil is required. Special areas shall include those listed in Regulation 10 of this Annex.

(11) "Instantaneous rate of discharge of oil content" means the rate of discharge of oil in litres per hour at any instant divided by the speed of the ship in knots at the same instant.

(12) "Tank" means an enclosed space which is formed by the permanent structure of a ship and which is designed for the carriage of liquid in bulk.

(13) "Wing tank" means any tank adjacent to the side shell plating.

(14) "Centre tank" means any tank inboard of a longitudinal bulkhead.

(15) "Slop tank" means a tank specifically designated for the collection of tank drainings, tank washings and other oily mixtures.

(16) "Clean ballast" means the ballast in a tank which since oil was last carried therein, has been so cleaned that effluent therefrom if it were discharged from a ship which is stationary into clean calm water on a clear day would not produce visible traces of oil on the surface of the water or on adjoining shorelines or cause

a sludge or emulsion to be deposited beneath the surface of the water or upon adjoining shorelines. If the ballast is discharged through an oil discharge monitoring and control system approved by the Administration, evidence based on such a system to the effect that the oil content of the effluent did not exceed 15 parts per million shall be determinative that the ballast was clean, notwithstanding the presence of visible traces.

(17) "Segregated ballast" means the ballast water introduced into a tank which is completely separated from the cargo oil and oil fuel system and which is permanently allocated to the carriage of ballast or to the carriage of ballast or cargoes other than oil or noxious substances as variously defined in the Annexes of the present Convention.

(18) "Length" (L) means 96 per cent of the total length on a waterline at 85 per cent of the least moulded depth measured from the top of the keel, or the length from the foreside of the stem to the axis of the rudder stock on that waterline, if that be greater. In ships designed with a rake of keel the waterline on which this length is measured shall be parallel to the designed waterline. The length (L) shall be measured in metres.

(19) "Forward and after perpendiculars" shall be taken at the forward and after ends of the length (L). The forward perpendicular shall coincide with the foreside of the stem on the waterline on which the length is measured.

(20) "Amidships" is at the middle of the length (L).

(21) "Breadth" (B) means the maximum breadth of the ship, measured amidships to the moulded line of the frame in a ship with a metal shell and to the outer surface of the hull in a ship with a shell of any other material. The breadth (B) shall be measured in metres.

(22) "Deadweight" (DW) means the difference in metric tons between the displacement of a ship in water of a specific gravity of 1.025 at the load waterline corresponding to the assigned summer freeboard and the lightweight of the ship.

(23) "Lightweight" means the displacement of a ship in metric tons without cargo, oil fuel, lubricating oil, ballast water, fresh water and feedwater in tanks, consumable stores, passengers and their effects.

(24) "Permeability" of a space means the ratio of the volume within that space which is assumed to be occupied by water to the total volume of that space.

(25) "Volumes" and "areas" in a ship shall be calculated in all cases to moulded lines.

Regulation 2

Application

(1)   Unless expressly provided otherwise, the provisions of this Annex shall apply to all ships.

(2) In ships other than oil tankers fitted with cargo spaces which are constructed and utilized to carry oil in bulk of an aggregate capacity of 200 cubic metres or more, the requirements of Regulations 9, 10, 14, 15(1), (2) and (3), 18, 20 and 24(4) of this Annex for oil tankers shall also apply to the construction and operation of those spaces, except that where such aggregate capacity is less than 1,000 cubic metres the requirements of Regulation 15(4) of this Annex may apply in lieu of Regulation 15(1), (2) and (3).

(3) Where a cargo subject to the provisions of Annex II of the present Convention is carried in a cargo space of an oil tanker, the appropriate requirements of Annex II of the present Convention shall also apply.

(4) (a) Any hydrofoil, air-cushion vehicle and other new type of vessel (near-surface craft, submarine craft, etc.) whose constructional features are such as to render the application of any of the provisions of Chapters II and III of this Annex relating to construction and equipment unreasonable or impracticable may be exempted by the Administration from such provisions, provided that the construction and equipment of that ship provides equivalent protection against pollution by oil, having regard to the service for which it is intended.

(b) Particulars of any such exemption granted by the Administration shall be indicated in the Certificate referred to in Regulation 5 of this Annex.

(c) The Administration which allows any such exemption shall, as soon as possible, but not more than ninety days thereafter, communicate to the Organization particulars of same and the reasons therefor, which the Organization shall circulate to the Parties to the Convention for their information and appropriate action, if any.

Regulation 3

Equivalents

(1) The Administration may allow any fitting, material, appliance or apparatus to be fitted in a ship as an alternative to that required by this Annex if such fitting, material, appliance or apparatus is at least as effective as that required by this Annex. This authority of the Administration shall not extend to substitution of operational methods to effect the control of discharge of oil as equivalent to those design and construction features which are prescribed by Regulations in this Annex.

(2) The Administration which allows a fitting, material, appliance or apparatus, as an alternative to that required by this Annex shall communicate to the Organization for circulation to the Parties to the Convention particulars thereof, for their information and appropriate action, if any.

Regulation 4

Surveys

(1) Every oil tanker of 150 tons gross tonnage and above, and every other ship of 400 tons gross tonnage and above shall be subject to the surveys specified below:

(a) An initial survey before the ship is put in service or before the Certificate required under Regulation 5 of this Annex is issued for the first time, which shall include a complete survey of its structure, equipment, fittings, arrangements and material in so far as the ship is covered by this Annex. This survey shall be such as to ensure that the structure, equipment, fittings, arrangements and material fully comply with the applicable requirements of this Annex.

(b) Periodical surveys at intervals specified by the Administration, but not exceeding five years, which shall be such as to ensure that the structure, equipment, fittings, arrangements and material fully comply with the applicable requirements of this Annex. However, where the duration of the International Oil Pollution Prevention Certificate (1973) is extended as specified in Regulation 8(3) or (4) of this Annex, the interval of the periodical survey may be extended correspondingly.

(c) Intermediate surveys at intervals specified by the Administration but not exceeding thirty months, which shall be such as to ensure that the equipment and associated pump and piping systems, including oil discharge monitoring and control systems, oily-water separating equipment and oil filtering systems, fully comply with the applicable requirements of this Annex and are in good working order. Such intermediate surveys shall be endorsed on the International Oil Pollution Prevention Certificate (1973) issued under Regulation 5 of this Annex.

(2) The Administration shall establish appropriate measures for ships which are not subject to the provisions of paragraph (1) of this Regulation in order to ensure that the applicable provisions of this Annex are complied with.

(3) Surveys of the ship as regards enforcement of the provisions of this Annex shall be carried out by officers of the Administration. The Administration may, however, entrust the surveys either to surveyors nominated for the purpose or to organizations recognized by it. In every case the Administration concerned fully guarantees the completeness and efficiency of the surveys.

(4) After any survey of the ship under this Regulation has been completed, no significant change shall be made in the structure, equipment, fittings, arrangements or material covered by the survey without the sanction of the Administration, except the direct replacement of such equipment or fittings.

Regulation 5

Issue of Certificate

(1) An International Oil Pollution Prevention Certificate (1973) shall be issued, after survey in accordance with the provisions of Regulation 4 of this Annex, to any oil tanker of 150 tons gross tonnage and above and any other ships of 400 tons gross tonnage and above which are engaged in voyages to ports or off-shore terminals under the jurisdiction of other Parties to the Convention. In the case of existing ships this requirement shall apply twelve months after the date of entry into force of the present Convention.

(2) Such Certificate shall be issued either by the Administration or by any persons or organization duly authorized by it. In every case the Administration assumes full responsibility for the Certificate.

Regulation 6

Issue of a Certificate by another Government

(1) The Government of a Party to the Convention may, at the request of the Administration, cause a ship to be surveyed and, if satisfied that the provisions of this Annex are complied with, shall issue or authorize the issue of an International Oil Pollution Prevention Certificate (1973) to the ship in accordance with this Annex.

(2) A copy of the Certificate and a copy of the survey report shall be transmitted as soon as possible to the requesting Administration.

(3) A Certificate so issued shall contain a statement to the effect that it has been issued at the request of the Administration and it shall have the same force and receive the same recognition as the Certificate issued under Regulation 5 of this Annex.

(4) No International Oil Pollution Prevention Certificate (1973) shall be issued to a ship which is entitled to fly the flag of a State which is not a Party.

Regulation 7

Form of Certificate

The International Oil Pollution Prevention Certificate (1973) shall be drawn up in an official language of the issuing country in the form corresponding to the model given in Appendix II to this Annex. If the language used is neither English nor French, the text shall include a translation into one of these languages.

Regulation 8

Duration of Certificate

(1) An International Oil Pollution Prevention Certificate (1973) shall be issued for a period specified by the Administration, which shall not exceed five years from the date of issue, except as provided in paragraphs (2), (3) and (4) of this Regulation.

(2) If a ship at the time when the Certificate expires is not in a port or off-shore terminal under the jurisdiction of the Party to the Convention whose flag the ship is entitled to fly, the Certificate may be extended by the Administration, but such extension shall be granted only for the purpose of allowing the ship to complete its voyage to the State whose flag the ship is entitled to fly or in which it is to be surveyed and then only in cases where it appears proper and reasonable to do so.

(3) No Certificate shall be thus extended for a period longer than five months and a ship to which such extension is granted shall not on its arrival in the State whose flag it is entitled to fly or the port in which it is to be surveyed, be entitled by virtue of such extension to leave that port or State without having obtained a new Certificate.

(4) A Certificate which has not been extended under the provisions of paragraph (2) of this Regulation may be extended by the Administration for a period of grace of up to one month from the date of expiry stated on it.

(5) A Certificate shall cease to be valid if significant alterations have taken place in the construction, equipment, fittings, arrangements, or material required without the sanction of the Administration, except the direct replacement of such equipment or fittings, or if intermediate surveys as specified by the Administration under Regulation 4(1)(c) of this Annex are not carried out.

(6) A Certificate issued to a ship shall cease to be valid upon transfer of such a ship to the flag of another State, except as provided in paragraph (7) of this Regulation.

(7) Upon transfer of a ship to the flag of another Party, the Certificate shall remain in force for a period not exceeding five months provided that it would not have expired before the end of that period, or until the Administration issues a replacement Certificate, whichever is earlier. As soon as possible after the transfer has taken place the Government of the Party whose flag the ship was formerly entitled to fly shall transmit to the Administration a copy of the Certificate carried by the ship before the transfer and, if available, a copy of the relevant survey report.

CHAPTER II — REQUIREMENTS FOR CONTROL OF OPERATIONAL POLLUTION

Regulation 9

Control of Discharge of Oil

(1) Subject to the provisions of Regulations 10 and 11 of this Annex and paragraph (2) of this Regulation, any discharge into the sea of oil or oily mixtures from ships to which this Annex applies shall be prohibited except when all the following conditions are satisfied:

(a) for an oil tanker, except as provided for in sub-paragraph (b) of this paragraph:

(i) the tanker is not within a special area;

(ii) the tanker is more than 50 nautical miles from the nearest land;

(iii) the tanker is proceeding en route;

(iv) the instantaneous rate of discharge of oil content does not exceed 60 litres per nautical mile;

(v) the total quantity of oil discharged into the sea does not exceed for existing tankers 1/15,000 of the total quantity of the particular cargo of which the residue formed a part, and for new tankers 1/30,000 of the total quantity of the particular cargo of which the residue formed a part; and

- (vi) the tanker has in operation, except as provided for in Regulation 15(3) of this Annex, an oil discharge monitoring and control system and a slop tank arrangement as required by Regulation 15 of this Annex;

(b) from a ship of 400 tons gross tonnage and above other than an oil tanker and from machinery space bilges excluding cargo pump room bilges of an oil tanker unless mixed with oil cargo residue:

- (i) the ship is not within a special area;
- (ii) the ship is more than 12 nautical miles from the nearest land;
- (iii) the ship is proceeding en route;
- (iv) the oil content of the effluent is less than 100 parts per million; and
- (v) the ship has in operation an oil discharge monitoring and control system, oily-water separating equipment, oil filtering system or other installation as required by Regulation 16 of this Annex.

(2) In the case of a ship of less than 400 tons gross tonnage other than an oil tanker whilst outside the special area, the Administration shall ensure that it is equipped as far as practicable and reasonable with installations to ensure the storage of oil residues on board and their discharge to reception facilities or into the sea in compliance with the requirements of paragraph (1)(b) of this Regulation.

(3) Whenever visible traces of oil are observed on or below the surface of the water in the immediate vicinity of a ship or its wake, Governments of Parties to the Convention should, to the extent they are reasonably able to do so, promptly investigate the facts bearing on the issue of whether there has been a violation of the provisions of this Regulation or Regulation 10 of this Annex. The investigation should include, in particular, the wind and sea conditions, the track and speed of the ship, other possible sources of the visible traces in the vicinity, and any relevant oil discharge records.

(4) The provisions of paragraph (1) of this Regulation shall not apply to the discharge of clean or segregated ballast. The provisions of sub-paragraph (1)(b) of this Regulation shall not apply to the discharge of oily mixture which without dilution has an oil content not exceeding 15 parts per million.

(5) No discharge into the sea shall contain chemicals or other substances in quantities or concentrations which are hazardous to the marine environment or chemicals or other substances introduced for the purpose of circumventing the conditions of discharge specified in this Regulation.

(6) The oil residues which cannot be discharged into the sea in compliance with paragraphs (1), (2) and (4) of this Regulation shall be retained on board or discharged to reception facilities.

Regulation 10

Methods for the Prevention of Oil Pollution from Ships while operating in Special Areas

(1) For the purposes of this Annex the special areas are the Mediterranean Sea area, the Baltic Sea area, the Black Sea area, the Red Sea area and the "Gulfs area" which are defined as follows:

- (a) The Mediterranean Sea area means the Mediterranean Sea proper including the gulfs and seas therein with the boundary between the Mediterranean and the Black Sea constituted by the 41°N parallel and bounded to the west by the Straits of Gibraltar at the meridian of 5°36'W.

- (b) The Baltic Sea area means the Baltic Sea proper with the Gulf of Bothnia, the Gulf of Finland and the entrance to the Baltic Sea bounded by the parallel of the Skaw in the Skagerrak at 57°44.8'N.

- (c) The Black Sea area means the Black Sea proper with the boundary between the Mediterranean and the Black Sea constituted by the parallel 41°N.

- (d) The Red Sea area means the Red Sea proper including the Gulfs of Suez and Aqaba bounded at the south by the rhumb line between Ras si Ane (12°8.5'N, 43°19.6'E) and Husn Murad (12°40.4'N, 43°30.2'E).

- (e) The Gulfs area means the sea area located north west of the rhumb line between Ras al Hadd (22°30'N, 59°48'E) and Ras Al Fasteh (25°04'N, 61°25'E).

(2)
- (a) Subject to the provisions of Regulation 11 of this Annex, any discharge into the sea of oil or oily mixture from any oil tanker and any ship of 400 tons gross tonnage and above other than an oil tanker shall be prohibited, while in a special area.

- (b) Such ships while in a special area shall retain on board all oil drainage and sludge, dirty ballast and tank washing waters and discharge them only to reception facilities.

(3)
- (a) Subject to the provisions of Regulation 11 of this Annex, any discharge into the sea of oil or oily mixture from a ship of less than 400 tons gross tonnage, other than an oil tanker, shall be prohibited while in a special area, except when the oil content of the effluent without dilution does not exceed 15 parts per million or alternatively when all of the following conditions are satisfied:

    (i) the ship is proceeding en route;

    (ii) the oil content of the effluent is less than 100 parts per million; and

    (iii) the discharge is made as far as practicable from the land, but in no case less than 12 nautical miles from the nearest land.

- (b) No discharge into the sea shall contain chemicals or other substances in quantities or concentrations which are hazardous to the marine environment or chemicals or other substances introduced for the purpose of circumventing the conditions of discharge specified in this Regulation.

- (c) The oil residues which cannot be discharged into the sea in compliance with sub-paragraph (a) of this paragraph shall be retained on board or discharged to reception facilities.

(4) The provisions of this Regulation shall not apply to the discharge of clean or segregated ballast.

(5) Nothing in this Regulation shall prohibit a ship on a voyage only part of which is in a special area from discharging outside the special area in accordance with Regulation 9 of this Annex.

(6) Whenever visible traces of oil are observed on or below the surface of the water in the immediate vicinity of a ship or its wake, the Governments of Parties to the Convention should, to the extent they are reasonably able to do so, promptly investigate the facts bearing on the issue of whether there has been a violation of the provisions of this Regulation or Regulation 9 of this Annex. The investigation should include, in particular, the wind and sea conditions, the track and speed of the ship, other possible sources of the visible traces in the vicinity, and any relevant oil discharge records.

(7) Reception facilities within special areas:

    (a) Mediterranean Sea, Black Sea and Baltic Sea areas:

        (i) The Government of each Party to the Convention, the coastline of which borders on any given special area undertakes to ensure that not later than 1 January 1977 all oil loading terminals and repair ports within the special area are provided with facilities adequate for the reception and treatment of all the dirty ballast and tank washing water from oil tankers. In addition all ports within the special area shall be provided with adequate reception facilities for other residues and oily mixtures from all ships. Such facilities shall have adequate capacity to meet the needs of the ships using them without causing undue delay.

        (ii) The Government of each Party having under its jurisdiction entrances to seawater courses with low depth contour which might require a reduction of draught by the discharge of ballast undertakes to ensure the provision of the facilities referred to in sub-paragraph (a)(i) of this paragraph but with the proviso that ships required to discharge slops or dirty ballast could be subject to some delay.

        (iii) During the period between the entry into force of the present Convention (if earlier than 1 January 1977) and 1 January 1977 ships while navigating in the special areas shall comply with the requirements of Regulation 9 of this Annex. However, the Governments of Parties the coastlines of which border any of the special areas under this sub-paragraph may establish a date earlier than 1 January 1977, but after the date of entry into force of the present Convention, from which the requirements of this Regulation in respect of the special areas in question shall take effect:

            (1) if all the reception facilities required have been provided by the date so established; and

            (2) provided that the Parties concerned notify the Organization of the date so established at least six months in advance, for circulation to other Parties.

        (iv) After 1 January 1977, or the date established in accordance with sub-paragraph (a)(iii) of this paragraph if earlier, each Party shall notify the Organization for transmission to the Contracting Governments concerned of all cases where the facilities are alleged to be inadequate.

(b) Red Sea area and Gulfs area:

(i) The Government of each Party the coastline of which borders on the special areas undertakes to ensure that as soon as possible all oil loading terminals and repair ports within these special areas are provided with facilities adequate for the reception and treatment of all the dirty ballast and tank washing water from tankers. In addition all ports within the special area shall be provided with adequate reception facilities for other residues and oily mixtures from all ships. Such facilities shall have adequate capacity to meet the needs of the ships using them without causing undue delay.

(ii) The Government of each Party having under its jurisdiction entrances to seawater courses with low depth contour which might require a reduction of draught by the discharge of ballast shall undertake to ensure the provision of the facilities referred to in sub-paragraph (b)(i) of this paragraph but with the proviso that ships required to discharge slops or dirty ballast could be subject to some delay.

(iii) Each Party concerned shall notify the Organization of the measures taken pursuant to provisions of sub-paragraph (b)(i) and (ii) of this paragraph. Upon receipt of sufficient notifications the Organization shall establish a date from which the requirements of this Regulation in respect of the area in question shall take effect. The Organization shall notify all Parties of the date so established no less than twelve months in advance of that date.

(iv) During the period between the entry into force of the present Convention and the date so established, ships while navigating in the special area shall comply with the requirements of Regulation 9 of this Annex.

(v) After such date oil tankers loading in ports in these special areas where such facilities are not yet available shall also fully comply with the requirements of this Regulation. However, oil tankers entering these special areas for the purpose of loading shall make every effort to enter the area with only clean ballast on board.

(vi) After the date on which the requirements for the special area in question take effect, each Party shall notify the Organization for transmission to the Parties concerned of all cases where the facilities are alleged to be inadequate.

(vii) At least the reception facilities as prescribed in Regulation 12 of this Annex shall be provided by 1 January 1977 or one year after the date of entry into force of the present Convention, whichever occurs later.

Regulation 11

Exceptions

Regulations 9 and 10 of this Annex shall not apply to:

(a) the discharge into the sea of oil or oily mixture necessary for the purpose of securing the safety of a ship or saving life at sea; or

(b) the discharge into the sea of oil or oily mixture resulting from damage to a ship or its equipment:

   (i) provided that all reasonable precautions have been taken after the occurrence of the damage or discovery of the discharge for the purpose of preventing or minimizing the discharge; and

   (ii) except if the owner or the Master acted either with intent to cause damage, or recklessly and with knowledge that damage would probably result; or

(c) the discharge into the sea of substances containing oil, approved by the Administration, when being used for the purpose of combating specific pollution incidents in order to minimize the damage from pollution. Any such discharge shall be subject to the approval of any Government in whose jurisdiction it is contemplated the discharge will occur.

Regulation 12

Reception Facilities

(1) Subject to the provisions of Regulation 10 of this Annex, the Government of each Party undertakes to ensure the provision at oil loading terminals, repair ports, and in other ports in which ships have oily residues to discharge, of facilities for the reception of such residues and oily mixtures as remain from oil tankers and other ships adequate to meet the needs of the ships using them without causing undue delay to ships.

(2) Reception facilities in accordance with paragraph (1) of this Regulation shall be provided in:

(a) all ports and terminals in which crude oil is loaded into oil tankers where such tankers have immediately prior to arrival completed a ballast voyage of not more than 72 hours or not more than 1,200 nautical miles;

(b) all ports and terminals in which oil other than crude oil in bulk is loaded at an average quantity of more than 1,000 metric tons per day;

(c) all ports having ship repair yards or tank cleaning facilities;

(d) all ports and terminals which handle ships provided with the sludge tank(s) required by Regulation 17 of this Annex;

(e) all ports in respect of oily bilge waters and other residues, which cannot be discharged in accordance with Regulation 9 of this Annex; and

(f) all loading ports for bulk cargoes in respect of oil residues from combination carriers which cannot be discharged in accordance with Regulation 9 of this Annex.

(3) The capacity for the reception facilities shall be as follows:

(a) Crude oil loading terminals shall have sufficient reception facilities to receive oil and oily mixtures which cannot be discharged in accordance with the provisions of Regulation 9(1)(a) of this Annex from all oil tankers on voyages as described in paragraph (2)(a) of this Regulation.

(b) Loading ports and terminals referred to in paragraph (2)(b) of this Regulation shall have sufficient reception facilities to receive oil and oily mixtures which cannot be discharged in accordance with the provisions of Regulation 9(1)(a) of this Annex from oil tankers which load oil other than crude oil in bulk.

(c) All ports having ship repair yards or tank cleaning facilities shall have sufficient reception facilities to receive all residues and oily mixtures which remain on board for disposal from ships prior to entering such yards or facilities.

(d) All facilities provided in ports and terminals under paragraph (2)(d) of this Regulation shall be sufficient to receive all residues retained according to Regulation 17 of this Annex from all ships that may reasonably be expected to call at such ports and terminals.

(e) All facilities provided in ports and terminals under this Regulation shall be sufficient to receive oily bilge waters and other residues which cannot be discharged in accordance with Regulation 9 of this Annex.

(f) The facilities provided in loading ports for bulk cargoes shall take into account the special problems of combination carriers as appropriate.

(4) The reception facilities prescribed in paragraphs (2) and (3) of this Regulation shall be made available no later than one year from the date of entry into force of the present Convention or by 1 January 1977, whichever occurs later.

(5) Each Party shall notify the Organization for transmission to the Parties concerned of all cases where the facilities provided under this Regulation are alleged to be inadequate.

Regulation 13

Segregated Ballast Oil Tankers

(1) Every new oil tanker of 70,000 tons deadweight and above shall be provided with segregated ballast tanks and shall comply with the requirements of this Regulation.

(2) The capacity of the segregated ballast tanks shall be so determined that the ship may operate safely on ballast voyages without recourse to the use of oil tanks for water ballast except as provided for in paragraph (3) of this Regulation. In all cases, however, the capacity of segregated ballast tanks shall be at least such that in

any ballast condition at any part of the voyage, including the conditions consisting of lightweight plus segregated ballast only, the ship's draughts and trim can meet each of the following requirements:

(a) the moulded draught amidships (dm) in metres (without taking into account any ship's deformation) shall not be less than:

$$dm = 2.0 + 0.02L;$$

(b) the draughts at the forward and after perpendiculars shall correspond to those determined by the draught amidships (dm), as specified in sub-paragraph (a) of this paragraph, in association with the trim by the stern of not greater than 0.015L; and

(c) in any case the draught at the after perpendicular shall not be less than that which is necessary to obtain full immersion of the propeller(s).

(3) In no case shall ballast water be carried in oil tanks except in weather conditions so severe that, in the opinion of the Master, it is necessary to carry additional ballast water in oil tanks for the safety of the ship. Such additional ballast water shall be processed and discharged in compliance with Regulation 9 and in accordance with the requirements of Regulation 15 of this Annex, and entry shall be made in the Oil Record Book referred to in Regulation 20 of this Annex.

(4) Any oil tanker which is not required to be provided with segregated ballast tanks in accordance with paragraph (1) of this Regulation may, however, be qualified as a segregated ballast tanker, provided that in the case of an oil tanker of 150 metres in length and above it fully complies with the requirements of paragraphs (2) and (3) of this Regulation and in the case of an oil tanker of less than 150 metres in length the segregated ballast conditions shall be to the satisfaction of the Administration.

Regulation 14

Segregation of Oil and Water Ballast

(1) Except as provided in paragraph (2) of this Regulation, in new ships of 4,000 tons gross tonnage and above other than oil tankers, and in new oil tankers of 150 tons gross tonnage and above, no ballast water shall be carried in any oil fuel tank.

(2) Where abnormal conditions or the need to carry large quantities of oil fuel render it necessary to carry ballast water which is not a clean ballast in any oil fuel tank, such ballast water shall be discharged to reception facilities or into the sea in compliance with Regulation 9 using the equipment specified in Regulation 16(2) of this Annex, and an entry shall be made in the Oil Record Book to this effect.

(3) All other ships shall comply with the requirements of paragraph (1) of this Regulation as far as reasonable and practicable.

Regulation 15

Retention of Oil on Board

(1) Subject to the provisions of paragraphs (5) and (6) of this Regulation, oil tankers of 150 tons gross tonnage and above shall be provided with arrangements in accordance with the requirements of paragraphs (2) and (3) of this Regulation, provided that in the case of existing tankers the requirements for oil discharge monitoring and control systems and slop tank arrangements shall apply three years after the date of entry into force of the present Convention.

(2) (a) Adequate means shall be provided for cleaning the cargo tanks and transferring the dirty ballast residue and tank washings from the cargo tanks into a slop tank approved by the Administration. In existing oil tankers, any cargo tank may be designated as a slop tank.

(b) In this system arrangements shall be provided to transfer the oily waste into a slop tank or combination of slop tanks in such a way that any effluent discharged into the sea will be such as to comply with the provisions of Regulation 9 of this Annex.

(c) The arrangements of the slop tank or combination of slop tanks shall have a capacity necessary to retain the slops generated by tank washing, oil residues and dirty ballast residues but the total shall be not less than 3 per cent of the oil carrying capacity of the ship, except that, where segregated ballast tanks are provided in accordance with Regulation 13 of this Annex, or where arrangements such as eductors involving the use of water additional to the washing water are not fitted, the Administration may accept 2 per cent. New oil tankers over 70,000 tons deadweight shall be provided with at least two slop tanks.

(d) Slop tanks shall be so designed particularly in respect of the position of inlets, outlets, baffles or weirs where fitted, so as to avoid excessive turbulence and entrainment of oil or emulsion with the water.

(3) (a) An oil discharge monitoring and control system approved by the Administration shall be fitted. In considering the design of the oil content meter to be incorporated in the system, the Administration shall have regard to the specification recommended by the Organization.* The system shall be fitted with a recording device to provide a continuous record of the discharge in litres per nautical mile and total quantity discharged, or the oil content and rate of discharge. This record shall be identifiable as to time and date and shall be kept for at least three years. The oil discharge monitor and control system shall come into operation when there is any discharge of effluent into the sea and shall be such as will ensure that any discharge of oily mixture is automatically stopped when the instantaneous rate of discharge of oil exceeds that permitted by Regulation 9(1)(a) of this Annex. Any failure of this monitoring and control system shall stop the discharge

---

* Reference is made to the Recommendation on International Performance Specifications for Oily-Water Separating Equipment and Oil Content Meters adopted by the Organization by Resolution A.233(VII).

and be noted in the Oil Record Book. A manually operated alternative method shall be provided and may be used in the event of such failure, but the defective unit shall be made operable before the oil tanker commences its next ballast voyage unless it is proceeding to a repair port. Existing oil tankers shall comply with all of the provisions specified above except that the stopping of the discharge may be performed manually and the rate of discharge may be estimated from the pump characteristic.

(b) Effective oil/water interface detectors approved by the Administration shall be provided for a rapid and accurate determination of the oil/water interface in slop tanks and shall be available for use in other tanks where the separation of oil and water is effected and from which it is intended to discharge effluent direct to the sea.

(c) Instructions as to the operation of the system shall be in accordance with an operational manual approved by the Administration. They shall cover manual as well as automatic operations and shall be intended to ensure that at no time shall oil be discharged except in compliance with the conditions specified in Regulation 9 of this Annex.*

(4) The requirements of paragraphs (1), (2) and (3) of this Regulation shall not apply to oil tankers of less than 150 tons gross tonnage, for which the control of discharge of oil under Regulation 9 of this Annex shall be effected by the retention of oil on board with subsequent discharge of all contaminated washings to reception facilities. The total quantity of oil and water used for washing and returned to a storage tank shall be recorded in the Oil Record Book. This total quantity shall be discharged to reception facilities unless adequate arrangements are made to ensure that any effluent which is allowed to be discharged into the sea is effectively monitored to ensure that the provisions of Regulation 9 of this Annex are complied with.

(5) The Administration may waive the requirements of paragraphs (1), (2) and (3) of this Regulation for any oil tanker which engages exclusively on voyages both of 72 hours or less in duration and within 50 miles from the nearest land, provided that the oil tanker is not required to hold and does not hold an International Oil Pollution Prevention Certificate (1973). Any such waiver shall be subject to the requirement that the oil tanker shall retain on board all oily mixtures for subsequent discharge to reception facilities and to the determination by the Administration that facilities available to receive such oily mixtures are adequate.

(6) Where in the view of the Organization equipment required by Regulation 9(1)(a)(vi) of this Annex and specified in sub-paragraph (3)(a) of this Regulation is not obtainable for the monitoring of discharge of light refined products (white oils), the Administration may waive compliance with such requirement, provided that discharge shall be permitted only in compliance with procedures established by the Organization which shall satisfy the conditions of Regulation 9(1)(a) of this Annex except the obligation to have an oil discharge monitoring and control system in operation. The Organization shall review the availability of equipment at intervals not exceeding twelve months.

---

* Reference is made to "Clean Seas Guide for Oil Tankers", published by the International Chamber of Shipping and the Oil Companies International Marine Forum.

(7) The requirements of paragraphs (1), (2) and (3) of this Regulation shall not apply to oil tankers carrying asphalt, for which the control of discharge of asphalt under Regulation 9 of this Annex shall be effected by the retention of asphalt residues on board with discharge of all contaminated washings to reception facilities.

Regulation 16

Oil Discharge Monitoring and Control System and Oily-Water Separating Equipment

(1) Any ship of 400 tons gross tonnage and above shall be fitted with an oily-water separating equipment or filtering system complying with the provisions of paragraph (6) of this Regulation. Any such ship which carries large quantities of oil fuel shall comply with paragraph 2 of this Regulation or paragraph (1) of Regulation 14.

(2) Any ship of 10,000 tons gross tonnage and above shall be fitted:

    (a) in addition to the requirements of paragraph (1) of this Regulation with an oil discharge monitoring and control system complying with paragraph (5) of this Regulation; or

    (b) as an alternative to the requirements of paragraph (1) and sub-paragraph (2)(a) of this Regulation, with an oily-water separating equipment complying with paragraph (6) of this Regulation and an effective filtering system, complying with paragraph (7) of this Regulation.

(3) The Administration shall ensure that ships of less than 400 tons gross tonnage are equipped, as far as practicable, to retain on board oil or oily mixtures or discharge them in accordance with the requirements of Regulation 9(1)(b) of this Annex.

(4) For existing ships the requirements of paragraphs (1), (2) and (3) of this Regulation shall apply three years after the date of entry into force of the present Convention.

(5) An oil discharge monitoring and control system shall be of a design approved by the Administration. In considering the design of the oil content meter to be incorporated into the system, the Administration shall have regard to the specification recommended by the Organization.* The system shall be fitted with a recording device to provide a continuous record of the oil content in parts per million. This record shall be identifiable as to time and date and shall be kept for at least three years. The monitoring and control system shall come into operation when there is any discharge of effluent into the sea and shall be such as will ensure that any discharge of oily mixture is automatically stopped when the oil content of effluent exceeds that permitted by Regulation 9(1)(b) of this Annex. Any failure of this monitoring and control system shall stop the discharge and be noted in the

---

* Reference is made to the Recommendation on International Performance Specifications for Oily-Water Separating Equipment and Oil Content Meters adopted by the Organization by Resolution A.233(VII).

Oil Record Book. The defective unit shall be made operable before the ship commences its next voyage unless it is proceeding to a repair port. Existing ships shall comply with all of the provisions specified above except that the stopping of the discharge may be performed manually.

(6) Oily-water separating equipment or an oil filtering system shall be of a design approved by the Administration and shall be such as will ensure that any oily mixture discharged into the sea after passing through the separator or filtering systems shall have an oil content of not more than 100 parts per million. In considering the design of such equipment, the Administration shall have regard to the specification recommended by the Organization.*

(7) The oil filtering system referred to in paragraph (2)(b) of this Regulation shall be of a design approved by the Administration and shall be such that it will accept the discharge from the separating system and produce an effluent the oil content of which does not exceed 15 parts per million. It shall be provided with alarm arrangements to indicate when this level cannot be maintained.

Regulation 17

Tanks for Oil Residues (Sludge)

(1) Every ship of 400 tons gross tonnage and above shall be provided with a tank or tanks of adequate capacity, having regard to the type of machinery and length of voyage, to receive the oily residues (sludges) which cannot be dealt with otherwise in accordance with the requirements of this Annex, such as those resulting from the purification of fuel and lubricating oils and oil leakages in the machinery spaces.

(2) In new ships, such tanks shall be designed and constructed so as to facilitate their cleaning and the discharge of residues to reception facilities. Existing ships shall comply with this requirement as far as is reasonable and practicable.

Regulation 18

Pumping, Piping and Discharge Arrangements of Oil Tankers

(1) In every oil tanker, a discharge manifold for connexion to reception facilities for the discharge of dirty ballast water or oil contaminated water shall be located on the open deck on both sides of the ship.

(2) In every oil tanker, pipelines for the discharge to the sea of effluent which may be permitted under Regulation 9 of this Annex shall be led to the open deck or to the ship's side above the waterline in the deepest ballast condition. Different piping arrangements to permit operation in the manner permitted in sub-paragraphs (4)(a) and (b) of this Regulation may be accepted.

---

* Reference is made to the Recommendation on International Performance Specifications for Oily-Water Separating Equipment and Oil Content Meters adopted by the Organization by Resolution A.233(VII).

(3) In new oil tankers means shall be provided for stopping the discharge of effluent into the sea from a position on upper deck or above located so that the manifold in use referred to in paragraph (1) of this Regulation and the effluent from the pipelines referred to in paragraph (2) of this Regulation may be visually observed. Means for stopping the discharge need not be provided at the observation position if a positive communication system such as telephone or radio system is provided between the observation position and the discharge control position.

(4) All discharges shall take place above the waterline except as follows:

    (a) Segregated ballast and clean ballast may be discharged below the waterline in ports or at offshore terminals.

    (b) Existing ships which, without modification, are not capable of discharging segregated ballast above the waterline may discharge segregated ballast below the waterline provided that an examination of the tank immediately before the discharge has established that no contamination with oil has taken place.

Regulation 19

Standard Discharge Connection

To enable pipes of reception facilities to be connected with the ship's discharge pipeline for residues from machinery bilges, both lines shall be fitted with a standard discharge connection in accordance with the following table:

STANDARD DIMENSIONS OF FLANGES FOR DISCHARGE CONNECTIONS

Description	Dimension
Outside diameter	215 mm
Inner diameter	According to pipe outside diameter
Bolt circle diameter	183 mm
Slots in flange	6 holes 22 mm in diameter equidistantly placed on a bolt circle of the above diameter, slotted to the flange periphery. The slot width to be 22 mm
Flange thickness	20 mm
Bolts and nuts: quantity, diameter	6, each of 20 mm in diameter and of suitable length

The flange is designed to accept pipes up to a maximum internal diameter of 125 mm and shall be of steel or other equivalent material having a flat face. This flange, together with a gasket of oilproof material, shall be suitable for a service pressure of 6 kg/cm^2.

Regulation 20

Oil Record Book

(1) Every oil tanker of 150 tons gross tonnage and above and every ship of 400 tons gross tonnage and above other than an oil tanker shall be provided with an Oil Record Book, whether as part of the ship's official log book or otherwise, in the form specified in Appendix III to this Annex.

(2) The Oil Record Book shall be completed on each occasion, on a tank-to-tank basis, whenever any of the following operations take place in the ship:

(a) For oil tankers

　　(i) loading of oil cargo;

　　(ii) internal transfer of oil cargo during voyage;

　　(iii) opening or closing before and after loading and unloading operations of valves or similar devices which inter-connect cargo tanks;

　　(iv) opening or closing of means of communication between cargo piping and seawater ballast piping;

　　(v) opening or closing of ships' side valves before, during and after loading and unloading operations;

　　(vi) unloading of oil cargo;

　　(vii) ballasting of cargo tanks;

　　(viii) cleaning of cargo tanks;

　　(ix) discharge of ballast except from segregated ballast tanks;

　　(x) discharge of water from slop tanks;

　　(xi) disposal of residues;

　　(xii) discharge overboard of bilge water which has accumulated in machinery spaces whilst in port, and the routine discharge at sea of bilge water which has accumulated in machinery spaces.

(b) For ships other than oil tankers

　　(i) ballasting or cleaning of fuel oil tanks or oil cargo spaces;

　　(ii) discharge of ballast or cleaning water from tanks referred to under (i) of this sub-paragraph;

　　(iii) disposal of residues;

　　(iv) discharge overboard of bilge water which has accumulated in machinery spaces whilst in port, and the routine discharge at sea of bilge water which has accumulated in machinery spaces.

(3) In the event of such discharge of oil or oily mixture as is referred to in Regulation 11 of this Annex or in the event of accidental or other exceptional discharge of oil not excepted by that Regulation, a statement shall be made in the Oil Record Book of the circumstances of, and the reasons for, the discharge.

(4) Each operation described in paragraph (2) of this Regulation shall be fully recorded without delay in the Oil Record Book so that all the entries in the book appropriate to that operation are completed. Each section of the book shall be signed by the officer or officers in charge of the operations concerned and shall be countersigned by the Master of the ship. The entries in the Oil Record Book shall be in an official language of the State whose flag the ship is entitled to fly, and, for ships holding an International Oil Pollution Prevention Certificate (1973), in English or French. The entries in an official national language of the State whose flag the ship is entitled to fly shall prevail in case of a dispute or discrepancy.

(5) The Oil Record Book shall be kept in such a place as to be readily available for inspection at all reasonable times and, except in the case of unmanned ships under tow, shall be kept on board the ship. It shall be preserved for a period of three years after the last entry has been made.

(6) The competent authority of the Government of a Party to the Convention may inspect the Oil Record Book on board any ship to which this Annex applies while the ship is in its port or offshore terminals and may make a copy of any entry in that book and may require the Master of the ship to certify that the copy is a true copy of such entry. Any copy so made which has been certified by the Master of the ship as a true copy of an entry in the ship's Oil Record Book shall be made admissible in any judicial proceedings as evidence of the facts stated in the entry. The inspection of an Oil Record Book and the taking of a certified copy by the competent authority under this paragraph shall be performed as expeditiously as possible without causing the ship to be unduly delayed.

Regulation 21

Special Requirements for Drilling Rigs and other Platforms

Fixed and floating drilling rigs when engaged in the exploration, exploitation and associated offshore processing of sea-bed mineral resources and other platforms shall comply with the requirements of this Annex applicable to ships of 400 tons gross tonnage and above other than oil tankers, except that:

(a) they shall be equipped as far as practicable with the installations required in Regulations 16 and 17 of this Annex;

(b) they shall keep a record of all operations involving oil or oily mixture discharges, in a form approved by the Administration; and

(c) in any special area and subject to the provisions of Regulation 11 of this Annex, the discharge into the sea of oil or oily mixture shall be prohibited except when the oil content of the discharge without dilution does not exceed 15 parts per million.

CHAPTER III — REQUIREMENTS FOR MINIMIZING OIL POLLUTION FROM OIL TANKERS DUE TO SIDE AND BOTTOM DAMAGES

Regulation 22

Damage Assumptions

(1) For the purpose of calculating hypothetical oil outflow from oil tankers, three dimensions of the extent of damage of a parallelepiped on the side and bottom of the ship are assumed as follows. In the case of bottom damages two conditions are set forth to be applied individually to the stated portions of the oil tanker.

(a) *Side damage*

    (i) Longitudinal extent (ℓ_c): $\frac{1}{3}L^{\frac{2}{3}}$ or 14.5 metres, whichever is less

    (ii) Transverse extent (t_c): $\frac{B}{5}$ or 11.5 metres, whichever is less
(inboard from the ship's side at right angles to the centreline at the level corresponding to the assigned summer freeboard)

    (iii) Vertical extent (v_c):   from the base line upwards without limit

(b) *Bottom damage*

	For 0.3L from the forward perpendicular of the ship	Any other part of the ship
(i) Longitudinal extent (ℓ_s):	$\frac{L}{10}$	$\frac{L}{10}$ or 5 metres, whichever is less
(ii) Transverse extent (t_s):	$\frac{B}{6}$ or 10 metres, whichever is less but not less than 5 metres	5 metres
(iii) Vertical extent from the base line (v_s):	$\frac{B}{15}$ or 6 metres, whichever is less	

(2) Wherever the symbols given in this Regulation appear in this Chapter, they have the meaning as defined in this Regulation.

Regulation 23

Hypothetical Outflow of Oil

(1) The hypothetical outflow of oil in the case of side damage (O_c) and bottom damage (O_s) shall be calculated by the following formulae with respect to compartments breached by damage to all conceivable locations along the length of the ship to the extent as defined in Regulation 22 of this Annex.

(a) for side damages:

$$O_c = \Sigma W_i + \Sigma K_i C_i \qquad (I)$$

(b) for bottom damages:

$$O_s = \tfrac{1}{3}(\Sigma Z_i W_i + \Sigma Z_i C_i) \qquad (II)$$

where: W_i = volume of a wing tank in cubic metres assumed to be breached by the damage as specified in Regulation 22 of this Annex; W_i for a segregated ballast tank may be taken equal to zero,

C_i = volume of a centre tank in cubic metres assumed to be breached by the damage as specified in Regulation 22 of this Annex; C_i for a segregated ballast tank may be taken equal to zero,

$K_i = 1 - \dfrac{b_i}{t_c}$ when b_i is equal to or greater than t_c, K_i shall be taken equal to zero,

$Z_i = 1 - \dfrac{h_i}{v_s}$ when h_i is equal to or greater than v_s, Z_i shall be taken equal to zero,

b_i = width of wing tank in metres under consideration measured inboard from the ship's side at right angles to the centreline at the level corresponding to the assigned summer freeboard,

h_i = minimum depth of the double bottom in metres under consideration; where no double bottom is fitted h_i shall be taken equal to zero.

Whenever symbols given in this paragraph appear in this Chapter, they have the meaning as defined in this Regulation.

(2) If a void space or segregated ballast tank of a length less than ℓ_c as defined in Regulation 22 of this Annex is located between wing oil tanks, O_c in formula (I) may be calculated on the basis of volume W_i being the actual volume of one such tank (where they are of equal capacity) or the smaller of the two tanks (if they differ in capacity) adjacent to such space, multiplied by S_i as defined below and taking for all other wing tanks involved in such a collision the value of the actual full volume.

$$S_i = 1 - \dfrac{\ell_i}{\ell_c}$$

where ℓ_i = length in metres of void space or segregated ballast tank under consideration.

(3) (a) Credit shall only be given in respect of double bottom tanks which are either empty or carrying clean water when cargo is carried in the tanks above.

(b) Where the double bottom does not extend for the full length and width of the tank involved, the double bottom is considered non-existent and the volume of the tanks above the area of the bottom damage shall be included in formula (II) even if the tank is not considered breached because of the installation of such a partial double bottom.

(c) Suction wells may be neglected in the determination of the value h_i provided such wells are not excessive in area and extend below the tank for a minimum distance and in no case more than half the height of the double bottom. If the depth of such a well exceeds half the height of the double bottom, h_i shall be taken equal to the double bottom height minus the well height.

Piping serving such wells if installed within the double bottom shall be fitted with valves or other closing arrangements located at the point of connexion to the tank served to prevent oil outflow in the event of damage to the piping. Such piping shall be installed as high from the bottom shell as possible. These valves shall be kept closed at sea at any time when the tank contains oil cargo, except that they may be opened only for cargo transfer needed for the purpose of trimming of the ship.

(4) In the case where bottom damage simultaneously involves four centre tanks, the value of O_s may be calculated according to the formula

$$O_s = \tfrac{1}{4}(\Sigma Z_i W_i + \Sigma Z_i C_i) \qquad (III)$$

(5) An Administration may credit as reducing oil outflow in case of bottom damage, an installed cargo transfer system having an emergency high suction in each cargo oil tank, capable of transferring from a breached tank or tanks to segregated ballast tanks or to available cargo tankage if it can be assured that such tanks will have sufficient ullage. Credit for such a system would be governed by ability to transfer in two hours of operation oil equal to one half of the largest of the breached tanks involved and by availability of equivalent receiving capacity in ballast or cargo tanks. The credit shall be confined to permitting calculation of O_s according to formula (III). The pipes for such suctions shall be installed at least at a height not less than the vertical extent of the bottom damage v_s. The Administration shall supply the Organization with the information concerning the arrangements accepted by it, for circulation to other Parties to the Convention.

Regulation 24

Limitation of Size and Arrangement of Cargo Tanks

(1) Every new oil tanker shall comply with the provision of this Regulation. Every existing oil tanker shall be required, within two years after the date of entry into force of the present Convention, to comply with the provisions of this Regulation if such a tanker falls into either of the following categories:

(a) a tanker, the delivery of which is after 1 January 1977; or

(b) a tanker to which both the following conditions apply:

    (i) delivery is not later than 1 January 1977; and

    (ii) the building contract is placed after 1 January 1974, or in cases where no building contract has previously been placed, the keel is laid or the tanker is at a similar stage of construction after 30 June 1974.

(2) Cargo tanks of oil tankers shall be of such size and arrangements that the hypothetical outflow O_c or O_s calculated in accordance with the provisions of Regulation 23 of this Annex anywhere in the length of the ship does not exceed 30,000 cubic metres or $400 \sqrt[3]{DW}$, whichever is the greater, but subject to a maximum of 40,000 cubic metres.

(3) The volume of any one wing cargo oil tank of an oil tanker shall not exceed seventy-five per cent of the limits of the hypothetical oil outflow referred to in paragraph (2) of this Regulation. The volume of any one centre cargo oil tank shall not exceed 50,000 cubic metres. However, in segregated ballast oil tankers as defined in Regulation 13 of this Annex, the permitted volume of a wing cargo oil tank situated between two segregated ballast tanks, each exceeding ℓ_c in length, may be increased to the maximum limit of hypothetical oil outflow provided that the width of the wing tanks exceeds t_c.

(4) The length of each cargo tank shall not exceed 10 metres or one of the following values, whichever is the greater:

(a) where no longitudinal bulkhead is provided:

    0.1L

(b) where a longitudinal bulkhead is provided at the centreline only:

    0.15L

(c) where two or more longitudinal bulkheads are provided:

    (i) for wing tanks:

        0.2L

    (ii) for centre tanks:

        (1) if $\frac{b_i}{B}$ is equal to or greater than $\frac{1}{5}$:

            0.2L

        (2) if $\frac{b_i}{B}$ is less than $\frac{1}{5}$:

            — where no centreline longitudinal bulkhead is provided:

                $(0.5 \frac{b_i}{B} + 0.1)L$

            — where a centreline longitudinal bulkhead is provided:

                $(0.25 \frac{b_i}{B} + 0.15)L$

(5) In order not to exceed the volume limits established by paragraphs (2), (3) and (4) of this Regulation and irrespective of the accepted type of cargo transfer system installed, when such system inter-connects two or more cargo tanks, valves or other similar closing devices shall be provided for separating the tanks from each other. These valves or devices shall be closed when the tanker is at sea.

(6) Lines of piping which run through cargo tanks in a position less than t_c from the ship's side or less than v_c from the ship's bottom shall be fitted with valves or similar closing devices at the point at which they open into any cargo tank. These valves shall be kept closed at sea at any time when the tanks contain cargo oil, except that they may be opened only for cargo transfer needed for the purpose of trimming of the ship.

Regulation 25

Subdivision and Stability

(1) Every new oil tanker shall comply with the subdivision and damage stability criteria as specified in paragraph (3) of this Regulation, after the assumed side or bottom damage as specified in paragraph (2) of this Regulation, for any operating draught reflecting actual partial or full load conditions consistent with trim and strength of the ship as well as specific gravities of the cargo. Such damage shall be applied to all conceivable locations along the length of the ship as follows:

    (a)   in tankers of more than 225 metres in length, anywhere in the ship's length;

    (b)   in tankers of more than 150 metres, but not exceeding 225 metres in length, anywhere in the ship's length except involving either after or forward bulkhead bounding the machinery space located aft. The machinery space shall be treated as a single floodable compartment;

    (c)   in tankers not exceeding 150 metres in length, anywhere in the ship's length between adjacent transverse bulkheads with the exception of the machinery space. For tankers of 100 metres or less in length where all requirements of paragraph (3) of this Regulation cannot be fulfilled without materially impairing the operational qualities of the ship, Administrations may allow relaxations from these requirements.

Ballast conditions where the tanker is not carrying oil in cargo tanks excluding any oil residues, shall not be considered.

(2) The following provisions regarding the extent and the character of the assumed damage shall apply:

    (a)   The extent of side or bottom damage shall be as specified in Regulation 22 of this Annex, except that the longitudinal extent of bottom damage within 0.3L from the forward perpendicular shall be the same as for side damage, as specified in Regulation 22(1)(a)(i) of this Annex. If any damage of lesser extent results in a more severe condition such damage shall be assumed.

(b) Where the damage involving transverse bulkheads is envisaged as specified in sub-paragraphs (1)(a) and (b) of this Regulation, transverse watertight bulkheads shall be spaced at least at a distance equal to the longitudinal extent of assumed damage specified in sub-paragraph (a) of this paragraph in order to be considered effective. Where transverse bulkheads are spaced at a lesser distance, one or more of these bulkheads within such extent of damage shall be assumed as non-existent for the purpose of determining flooded compartments.

(c) Where the damage between adjacent transverse watertight bulkheads is envisaged as specified in sub-paragraph (1)(c) of this Regulation, no main transverse bulkhead or a transverse bulkhead bounding side tanks or double bottom tanks shall be assumed damaged, unless:

   (i) the spacing of the adjacent bulkheads is less than the longitudinal extent of assumed damage specified in sub-paragraph (a) of this paragraph; or

   (ii) there is a step or a recess in a transverse bulkhead of more than 3.05 metres in length, located within the extent of penetration of assumed damage. The step formed by the after peak bulkhead and after peak tank top shall not be regarded as a step for the purpose of this Regulation.

(d) If pipes, ducts or tunnels are situated within the assumed extent of damage, arrangements shall be made so that progressive flooding cannot thereby extend to compartments other than those assumed to be floodable for each case of damage.

(3) Oil tankers shall be regarded as complying with the damage stability criteria if the following requirements are met:

(a) The final waterline, taking into account sinkage, heel and trim, shall be below the lower edge of any opening through which progressive flooding may take place. Such openings shall include air pipes and those which are closed by means of weathertight doors or hatch covers and may exclude those openings closed by means of watertight manhole covers and flush scuttles, small watertight cargo tank hatch covers which maintain the high integrity of the deck, remotely operated watertight sliding doors, and side scuttles of the non-opening type.

(b) In the final stage of flooding, the angle of heel due to unsymmetrical flooding shall not exceed 25 degrees, provided that this angle may be increased up to 30 degrees if no deck edge immersion occurs.

(c) The stability in the final stage of flooding shall be investigated and may be regarded as sufficient if the righting lever curve has at least a range of 20 degrees beyond the position of equilibrium in association with a maximum residual righting lever of at least 0.1 metre. The Administration shall give consideration to the potential hazard presented by protected or unprotected openings which may become temporarily immersed within the range of residual stability.

(d) The Administration shall be satisfied that the stability is sufficient during intermediate stages of flooding.

(4) The requirements of paragraph (1) of this Regulation shall be confirmed by calculations which take into consideration the design characteristics of the ship, the arrangements, configuration and contents of the damaged compartments; and the distribution, specific gravities and the free surface effect of liquids. The calculations shall be based on the following:

    (a) Account shall be taken of any empty or partially filled tank, the specific gravity of cargoes carried, as well as any outflow of liquids from damaged compartments.

    (b) The permeabilities are assumed as follows:

Spaces	Permeability
Appropriated to stores	0.60
Occupied by accommodation	0.95
Occupied by machinery	0.85
Voids	0.95
Intended for consumable liquids	0 or 0.95*
Intended for other liquids	0 to 0.95**

    * Whichever results in the more severe requirements.

    ** The permeability of partially filled compartments shall be consistent with the amount of liquid carried.

    (c) The buoyancy of any superstructure directly above the side damage shall be disregarded. The unflooded parts of superstructures beyond the extent of damage, however, may be taken into consideration provided that they are separated from the damaged space by watertight bulkheads and the requirements of sub-paragraph (3)(a) of this Regulation in respect of these intact spaces are complied with. Hinged watertight doors may be acceptable in watertight bulkheads in the superstructure.

    (d) The free surface effect shall be calculated at an angle of heel of 5 degrees for each individual compartment. The Administration may require or allow the free surface corrections to be calculated at an angle of heel greater than 5 degrees for partially filled tanks.

    (e) In calculating the effect of free surfaces of consumable liquids it shall be assumed that, for each type of liquid at least one transverse pair or a single centreline tank has a free surface and the tank or combination of tanks to be taken into account shall be those where the effect of free surfaces is the greatest.

(5) The Master of every oil tanker and the person in charge of a non-self-propelled oil tanker to which this Annex applies shall be supplied in an approved form with:

    (a) information relative to loading and distribution of cargo necessary to ensure compliance with the provisions of this Regulation; and

    (b) data on the ability of the ship to comply with damage stability criteria as determined by this Regulation, including the effect of relaxations that may have been allowed under sub-paragraph (1)(c) of this Regulation.

Appendix I

LIST OF OILS*

Asphalt solutions
Blending Stocks
Roofers Flux
Straight Run Residue

Oils
Clarified
Crude Oil
Mixtures containing crude oil
Diesel Oil
Fuel Oil No.4
Fuel Oil No.5
Fuel Oil No.6
Residual Fuel Oil
Road Oil
Transformer Oil
Aromatic Oil (excluding vegetable oil)
Lubricating Oils and Blending Stocks
Mineral Oil
Motor Oil
Penetrating Oil
Spindle Oil
Turbine Oil

Distillates
Straight Run
Flashed Feed Stocks

Gas Oil
Cracked

Gasoline Blending Stocks
Alkylates – fuel
Reformates
Polymer – fuel

Gasolines
Casinghead (natural)
Automotive
Aviation
Straight Run
Fuel Oil No.1 (Kerosene)
Fuel Oil No.1-D
Fuel Oil No.2
Fuel Oil No.2-D

Jet Fuels
JP-1 (Kerosene)
JP-3
JP-4
JP-5 (Kerosene, Heavy)
Turbo Fuel
Kerosene
Mineral Spirit

Naphtha
Solvent
Petroleum
Heartcut Distillate Oil

---

* The list of oils shall not necessarily be considered as comprehensive.

Appendix II

FORM OF CERTIFICATE

INTERNATIONAL OIL POLLUTION PREVENTION CERTIFICATE (1973)

Issued under the Provisions of the International Convention for the Prevention of Pollution from Ships, 1973, under the Authority of the Government of

....
(full designation of the country)

by ....
(full designation of the competent person or organization authorized under the provisions of the International Convention for the Prevention of Pollution from Ships, 1973)

Name of Ship	Distinctive Number or Letter	Port of Registry	Gross Tonnage

Type of ship:

    Oil tanker, including combination carrier*

    Asphalt carrier*

    Ship other than an oil tanker with cargo tanks coming under Regulation 2(2) of Annex I of the Convention*

    Ship other than any of the above*

New/existing ship*

Date of building or major conversion contract................................

Date on which keel was laid or ship was at a similar stage of construction or on which major conversion was commenced....................

Date of delivery or completion of major conversion..........................

---

* Delete as appropriate.

PART A   ALL SHIPS

The ship is equipped with:

for ships of 400 tons gross tonnage and above:

(a) oily-water separating equipment* (capable of producing the effluent with an oil content not exceeding 100 parts per million) or

(b) an oil filtering system* (capable of producing the effluent with an oil content not exceeding 100 parts per million)

for ships of 10,000 tons gross tonnage and above:

(c) an oil discharge monitoring and control system* (additional to (a) or (b) above) or

(d) oily-water separating equipment and an oil filtering system* (capable of producing the effluent with an oil content not exceeding 15 parts per million) in lieu of (a) or (b) above.

Particulars of requirements from which exemption is granted under Regulation 2(2) and 2(4)(a) of Annex I of the Convention:

....

....

Remarks:

---

* Delete as appropriate.

PART B   OIL TANKER[1] [2]

Deadweight .... metric tons. Length of ship .... metres.

It is certified that this ship is:

(a) required to be constructed according to and complies with[3]

(b) not required to be constructed according to[3]

(c) not required to be constructed according to, but complies with[3]

the requirements of Regulation 24 of Annex I of the Convention.

The capacity of segregated ballast tanks is .... cubic metres and complies with the requirements of Regulation 13 of Annex I of the Convention.

The segregated ballast is distributed as follows:

Tank	Quantity	Tank	Quantity

---

[1] This Part should be completed for oil tankers including combination carriers and asphalt carriers, and those entries which are applicable should be completed for ships other than oil tankers which are constructed and utilized to carry oil in bulk of an aggregate capacity of 200 cubic metres or above.

[2] This page need not be reproduced on a Certificate issued to any ship other than those referred to in footnote 1.

[3] Delete as appropriate.

THIS IS TO CERTIFY:

That the ship has been surveyed in accordance with Regulation 4 of Annex I of the International Convention for the Prevention of Pollution from Ships, 1973, concerning the prevention of pollution by oil; and

That the survey shows that the structure, equipment, fittings, arrangement and material of the ship and the condition thereof are in all respects satisfactory and that the ship complies with the applicable requirements of Annex I of the Convention.

This Certificate is valid until ....
subject to intermediate survey(s) at intervals of ....

Issued at..
(place of issue of Certificate)

........................19..   ....
          (Signature of duly authorized official issuing the Certificate)

(Seal or stamp of the issuing Authority, as appropriate)

Endorsement for existing ships[4]

This is to certify that this ship has been so equipped as to comply with the requirements of the International Convention for the Prevention of Pollution from Ships, 1973 as relating to existing ships three years from the date of entry into force of the Convention.

        Signed ....
        (Signature of duly authorized official)

        Place of endorsement ....

        Date of endorsement.................

(Seal or stamp of the Authority, as appropriate)

---

[4] This entry need not be reproduced on a Certificate other than the first Certificate issued to any ship.

Intermediate survey

This is to certify that at an intermediate survey required by Regulation 4(1)(c) of Annex I of the Convention, this ship and the condition thereof are found to comply with the relevant provisions of the Convention.

Signed ....
(Signature of duly authorized official)

Place ....

Date ....

(Seal or stamp of the Authority, as appropriate)

Signed ....
(Signature of duly authorized official)

Place ....

Date ....

(Seal or stamp of the Authority, as appropriate)

Under the provisions of Regulation 8(2) and (4) of Annex I of the Convention the validity of this Certificate is extended until

....

Signed ....
(Signature of duly authorized official)

Place ....

Date ....

(Seal or stamp of the Authority, as appropriate)

Appendix III

FORM OF OIL RECORD BOOK

OIL RECORD BOOK

I – FOR OIL TANKERS[1]

Name of ship ....

Total cargo carrying capacity of ship in cubic metres ....

Voyage from ....(date)....... to..............(date)......

(a) Loading of oil cargo

1. Date and place of loading			
2. Types of oil loaded			
3. Identity of tank(s) loaded			
4. Closing of applicable cargo tank valves and applicable line cut-off valves on completion of loading[2]			

The undersigned certifies that in addition to the above, all sea valves, overboard discharge valves, cargo tank and pipeline connections and inter-connections, were secured on completion of loading oil cargo.

Date of entry................... Officer in charge...................

Master ....

---

[1] This Part should be completed for oil tankers including combination carriers and asphalt carriers, and those entries which are applicable shall be completed for ships other than oil tankers which are constructed and utilized to carry oil in bulk of an aggregate capacity of 200 cubic metres or above. This Part need not be reproduced on an Oil Record Book issued to any ship other than those referred to above.

[2] Applicable valves and similar devices are those referred to in Regulations 20(2)(a)(iii), 23 and 24 of Annex I of the Convention.

(b) Internal transfer of oil cargo during voyage

5. Date of internal transfer					
6. Identity of tank(s)	(i)	From			
	(ii)	To			
7. Was(were) tank(s) in 6(i) emptied?					

The undersigned certifies that in addition to the above, all sea valves, overboard discharge valves, cargo tank and pipeline connections and inter-connections, were secured on completion of internal transfer of oil cargo.

Date of entry.................... Officer in charge.....................

Master............................

(c) Unloading of oil cargo

8. Date and place of unloading			
9. Identity of tank(s) unloaded			
10. Was(were) tank(s) emptied?			
11. Opening of applicable cargo tank valves and applicable line cut-off valves prior to cargo unloading[2]			
12. Closing of applicable cargo tank valves and applicable line cut-off valves on completion of unloading[2]			

The undersigned certifies that in addition to the above, all sea valves, overboard discharge valves, cargo tank and pipeline connections and inter-connections, were secured on completion of unloading of oil cargo.

Date of entry.................... Officer in charge.....................

Master............................

(d) Ballasting of cargo tanks

13. Identity of tank(s) ballasted			
14. Date and position of ship at start of ballasting			
15. If valves connecting cargo lines and segregated ballast lines were used give time, date and position of ship when valves were (a) opened, and (b) closed			

The undersigned certifies that in addition to the above all sea valves, overboard discharge valves, cargo tank and pipeline connections and inter-connections, were secured on completion of ballasting.

Date of entry.................... Officer in charge....................

Master............................

(e) Cleaning of cargo tanks

16. Identity of tank(s) cleaned			
17. Date and duration of cleaning			
18. Methods of cleaning[3]			

Date of entry.................... Officer in charge....................

Master............................

---

[3] Hand hosing, machine washing and/or chemical cleaning. Where chemically cleaned, the chemical concerned and the amount used should be stated.

72

(f) Discharge of dirty ballast

19. Identity of tank(s)			
20. Date and position of ship at start of discharge to sea			
21. Date and position of ship at finish of discharge to sea			
22. Ship's speed(s) during discharge			
23. Quantity discharged to sea			
24. Quantity of polluted water transferred to slop tank(s) (identify slop tank(s))			
25. Date and port of discharge into shore reception facilities (if applicable)			
26. Was any part of the discharge conducted during darkness, if so, for how long?			
27. Was a regular check kept on the effluent and the surface of the water in the locality of the discharge?			
28. Was any oil observed on the surface of the water in the locality of the discharge?			

Date of entry.................... Officer in charge....................

Master............................

(g) Discharge of water from slop tanks

29. Identity of slop tank(s)			
30. Time of settling from last entry of residues, or			
31. Time of settling from last discharge			
32. Date, time and position of ship at start of discharge			
33. Sounding of total contents at start of discharge			
34. Sounding of oil/water interface at start of discharge			
35. Bulk quantity discharged and rate of discharge			
36. Final quantity discharged and rate of discharge			
37. Date, time and position of ship at end of discharge			
38. Ship's speed(s) during discharge			
39. Sounding of oil/water interface at end of discharge			
40. Was any part of the discharge conducted during darkness, if so, for how long?			
41. Was a regular check kept on the effluent and the surface of the water in the locality of the discharge?			
42. Was any oil observed on the surface of the water in the locality of the discharge?			

Date of entry..................... Officer in charge......................

Master............................

(h) Disposal of residues

43. Identity of tank(s)			
44. Quantity disposed from each tank			
45. Method of disposal of residue: (a) Reception facilities (b) Mixed with cargo (c) Transferred to another (other) tank(s) (identify tank(s)) (d) Other method (state which)			
46. Date and port of disposal of residue			

Date of entry.................... Officer in charge.....................

Master............................

(i) Discharge of clean ballast contained in cargo tanks

47. Date and position of ship at commencement of discharge of clean ballast			
48. Identity of tank(s) discharged			
49. Was(were) the tank(s) empty on completion?			
50. Position of vessel on completion if different from 47			
51. Was any part of the discharge conducted during darkness, if so, for how long?			
52. Was a regular check kept on the effluent and the surface of the water in the locality of the discharge?			
53. Was any oil observed on the surface of the water in the locality of the discharge?			

Date of entry.................... Officer in charge....................

Master............................

(j) Discharge overboard of bilge water containing oil which has accumulated in machinery spaces whilst in port[4]

54. Port			
55. Duration of stay			
56. Quantity disposed			
57. Date and place of disposal			
58. Method of disposal (state whether a separator was used)			

Date of entry.................... Officer in charge....................

Master............................

(k) Accidental or other exceptional discharges of oil

59. Date and time of occurrence			
60. Place or position of ship at time of occurrence			
61. Approximate quantity and type of oil			
62. Circumstances of discharge or escape, the reasons therefor and general remarks			

Date of entry.................... Officer in charge....................

Master............................

---

[4] Where the pump starts automatically and discharges through a separator at all times it will be sufficient to enter each day "Automatic discharge from bilges through a separator".

(1) Has the oil monitoring and control system been out of operation at any time when discharging overboard? If so, give time and date of failure and time and date of restoration and confirm that this was due to equipment failure and state reason if known ....

....

....

Date of entry .... Officer in charge ....

Master ....

(m) Additional operational procedures and general remarks ....

....

....

....

For oil tankers of less than 150 tons gross tonnage operating in accordance with Regulation 15(4) of Annex I of the Convention, an appropriate oil record book should be developed by the Administration.

For asphalt carriers, a separate oil record book may be developed by the Administration utilizing sections (a), (b), (c), (e), (h), (j), (k) and (m) of this form of oil record book.

II – FOR SHIPS OTHER THAN OIL TANKERS

Name of ship ....

Operations from .... (date), to .... (date)

(a) Ballasting or cleaning of oil fuel tanks

1. Identity of tank(s) ballasted			
2. Whether cleaned since they last contained oil and, if not, type of oil previously carried			
3. Date and position of ship at start of cleaning			
4. Date and position of ship at start of ballasting			

Date of entry .... Officer in charge ....

Master ....

(b) Discharge of dirty ballast or cleaning water from tanks referred to under section (a)

5. Identity of tank(s)			
6. Date and position of ship at start of discharge			
7. Date and position of ship at finish of discharge			
8. Ship's speed(s) during discharge			
9. Method of discharge (state whether to reception facility or through installed equipment)			
10. Quantity discharged			

Date of entry .... Officer in charge ....

Master ....

(c) Disposal of residues

11. Quantity of residue retained on board			

12. Methods of disposal of residue:
    (a) reception facilities
    (b) mixed with next bunkering
    (c) transferred to another (other) tank
    (d) other method (state which)			

13. Date and port of disposal of residue			

Date of entry.................... Officer in charge....................

Master ....

(d) Discharge overboard of bilge water containing oil which has accumulated in machinery spaces whilst in port[5]

14. Port			
15. Duration of stay			
16. Quantity discharged			
17. Date and place of discharge			

18. Method of discharge:
    (a) through oily-water separating equipment;
    (b) through oil filtering system;
    (c) through oily-water separating equipment and an oil filtering system;
    (d) to reception facilities			

Date of entry.................... Officer in charge....................

Master ....

---

[5] Where the pump starts automatically and discharges through a separator at all times it will be sufficient to enter each day "Automatic discharge from bilges through a separator".

(e) Accidental or other exceptional discharges of oil

19. Date and time of occurrence				
20. Place or position of ship at time of occurrence				
21. Approximate quantity and type of oil				
22. Circumstances of discharge or escape, the reasons therefor and general remarks				

Date of entry..................... Officer in charge.....................

Master ....

(f) Has the required oil monitoring and control system been out of operation at any time when discharging overboard? If so, state time and date of failure and time and date of restoration, and confirm that this was due to equipment failure, and state reason if known

Date of entry..................... Officer in charge.....................

Master ....

(g) New ships of 4,000 tons gross tonnage and above: has dirty ballast been carried in oil fuel tanks?
Yes/No.............

If so, state which tanks were so ballasted and method of discharge of the dirty ballast..
....
....

Date of entry..................... Officer in charge.....................

Master ....

(h) Additional operational procedures and general remarks ....
....
....
....

Date of entry..................... Officer in charge.....................

Master ....

ANNEX II

REGULATIONS FOR THE CONTROL OF POLLUTION BY NOXIOUS LIQUID SUBSTANCES IN BULK

Regulation 1

Definitions

For the purposes of this Annex:

(1) "Chemical tanker" means a ship constructed or adapted primarily to carry a cargo of noxious liquid substances in bulk and includes an "oil tanker" as defined in Annex I of the present Convention when carrying a cargo or part cargo of noxious liquid substances in bulk.

(2) "Clean ballast" means ballast carried in a tank which, since it was last used to carry a cargo containing a substance in Category A, B, C or D has been thoroughly cleaned and the residues resulting therefrom have been discharged and the tank emptied in accordance with the appropriate requirements of this Annex.

(3) "Segregated ballast" means ballast water introduced into a tank permanently allocated to the carriage of ballast or to the carriage of ballast or cargoes other than oil or noxious liquid substances as variously defined in the Annexes of the present Convention, and which is completely separated from the cargo and oil fuel system.

(4) "Nearest land" is as defined in Regulation 1(9) of Annex I of the present Convention.

(5) "Liquid substances" are those having a vapour pressure not exceeding 2.8 kp/cm^2 at a temperature of 37.8°C.

(6) "Noxious liquid substance" means any substance designated in Appendix II to this Annex or provisionally assessed under the provisions of Regulation 3(4) as falling into Category A, B, C or D.

(7) "Special area" means a sea area where for recognized technical reasons in relation to its oceanographic and ecological condition and to its peculiar transportation traffic the adoption of special mandatory methods for the prevention of sea pollution by noxious liquid substances is required.

Special areas shall be:

(a) The Baltic Sea Area, and

(b) The Black Sea Area.

(8) "Baltic Sea Area" is as defined in Regulation 10(1)(b) of Annex I of the present Convention.

(9) "Black Sea Area" is as defined in Regulation 10(1)(c) of Annex I of the present Convention.

Regulation 2

Application

(1)   Unless expressly provided otherwise the provisions of this Annex shall apply to all ships carrying noxious liquid substances in bulk.

(2)   Where a cargo subject to the provisions of Annex I of the present Convention is carried in a cargo space of a chemical tanker, the appropriate requirements of Annex I of the present Convention shall also apply.

(3)   Regulation 13 of this Annex shall apply only to ships carrying substances which are categorized for discharge control purposes in Category A, B or C.

Regulation 3

Categorization and Listing of Noxious Liquid Substances

(1)   For the purpose of the Regulations of this Annex, except Regulation 13, noxious liquid substances shall be divided into four categories as follows:

(a)   Category A — Noxious liquid substances which if discharged into the sea from tank cleaning or deballasting operations would present a major hazard to either marine resources or human health or cause serious harm to amenities or other legitimate uses of the sea and therefore justify the application of stringent anti-pollution measures.

(b)   Category B — Noxious liquid substances which if discharged into the sea from tank cleaning or deballasting operations would present a hazard to either marine resources or human health or cause harm to amenities or other legitimate uses of the sea and therefore justify the application of special anti-pollution measures.

(c)   Category C — Noxious liquid substances which if discharged into the sea from tank cleaning or deballasting operations would present a minor hazard to either marine resources or human health or cause minor harm to amenities or other legitimate uses of the sea and therefore require special operational conditions.

(d)   Category D — Noxious liquid substances which if discharged into the sea from tank cleaning or deballasting operations would present a recognizable hazard to either marine resources or human health or cause minimal harm to amenities or other legitimate uses of the sea and therefore require some attention in operational conditions.

(2)   Guidelines for use in the categorization of noxious liquid substances are given in Appendix I to this Annex.

(3)   The list of noxious liquid substances carried in bulk and presently categorized which are subject to the provisions of this Annex is set out in Appendix II to this Annex.

(4) Where it is proposed to carry a liquid substance in bulk which has not been categorized under paragraph (1) of this Regulation or evaluated as referred to in Regulation 4(1) of this Annex, the Governments of Parties to the Convention involved in the proposed operation shall establish and agree on a provisional assessment for the proposed operation on the basis of the guidelines referred to in paragraph (2) of this Regulation. Until full agreement between the Governments involved has been reached, the substance shall be carried under the most severe conditions proposed. As soon as possible, but not later than ninety days after its first carriage, the Administration concerned shall notify the Organization and provide details of the substance and the provisional assessment for prompt circulation to all Parties for their information and consideration. The Government of each Party shall have a period of ninety days in which to forward its comments to the Organization, with a view to the assessment of the substance.

Regulation 4

Other Liquid Substances

(1) The substances listed in Appendix III to this Annex have been evaluated and found to fall outside the Categories A, B, C and D, as defined in Regulation 3(1) of this Annex because they are presently considered to present no harm to human health, marine resources, amenities or other legitimate uses of the sea, when discharged into the sea from tank cleaning or deballasting operations.

(2) The discharge of bilge or ballast water or other residues or mixtures containing only substances listed in Appendix III to this Annex shall not be subject to any requirement of this Annex.

(3) The discharge into the sea of clean ballast or segregated ballast shall not be subject to any requirement of this Annex.

Regulation 5

Discharge of Noxious Liquid Substances

Categories A, B and C Substances outside Special Areas and Category D Substances in all Areas

Subject to the provisions of Regulation 6 of this Annex,

(1) The discharge into the sea of substances in Category A as defined in Regulation 3(1)(a) of this Annex or of those provisionally assessed as such or ballast water, tank washings, or other residues or mixtures containing such substances shall be prohibited. If tanks containing such substances or mixtures are to be washed, the resulting residues shall be discharged to a reception facility until the concentration of the substance in the effluent to such facility is at or below the residual concentration prescribed for that substance in column III of Appendix II to this Annex and until the tank is empty. Provided that the residue then remaining in the tank is subsequently diluted by the addition of a volume of water of not less than 5 per cent of the total volume of the tank, it may be discharged into the sea when all the following conditions are also satisfied:

(a) the ship is proceeding en route at a speed of at least 7 knots in the case of self-propelled ships or at least 4 knots in the case of ships which are not self-propelled;

(b) the discharge is made below the waterline, taking into account the location of the seawater intakes; and

(c) the discharge is made at a distance of not less than 12 nautical miles from the nearest land and in a depth of water of not less than 25 metres.

(2) The discharge into the sea of substances in Category B as defined in Regulation 3(1)(b) of this Annex or of those provisionally assessed as such, or ballast water, tank washings, or other residues or mixtures containing such substances shall be prohibited except when all the following conditions are satisfied:

(a) the ship is proceeding en route at a speed of at least 7 knots in the case of self-propelled ships or at least 4 knots in the case of ships which are not self-propelled;

(b) the procedures and arrangements for discharge are approved by the Administration. Such procedures and arrangements shall be based upon standards developed by the Organization and shall ensure that the concentration and rate of discharge of the effluent is such that the concentration of the substance in the wake astern of the ship does not exceed 1 part per million;

(c) the maximum quantity of cargo discharged from each tank and its associated piping system does not exceed the maximum quantity approved in accordance with the procedures referred to in sub-paragraph (b) of this paragraph, which shall in no case exceed the greater of 1 cubic metre or 1/3,000 of the tank capacity in cubic metres;

(d) the discharge is made below the waterline, taking into account the location of the seawater intakes; and

(e) the discharge is made at a distance of not less than 12 nautical miles from the nearest land and in a depth of water of not less than 25 metres.

(3) The discharge into the sea of substances in Category C as defined in Regulation 3(1)(c) of this Annex or of those provisionally assessed as such, or ballast water, tank washings, or other residues or mixtures containing such substances shall be prohibited except when all the following conditions are satisfied:

(a) the ship is proceeding en route at a speed of at least 7 knots in the case of self-propelled ships or at least 4 knots in the case of ships which are not self-propelled;

(b) the procedures and arrangements for discharge are approved by the Administration. Such procedures and arrangements shall be based upon standards developed by the Organization and shall ensure that the concentration and rate of discharge of the effluent is such that the concentration of the substance in the wake astern of the ship does not exceed 10 parts per million;

(c) the maximum quantity of cargo discharged from each tank and its associated piping system does not exceed the maximum quantity approved in accordance with the procedures referred to in sub-paragraph (b) of this paragraph, which shall in no case exceed the greater of 3 cubic metres or 1/1,000 of the tank capacity in cubic metres;

(d) the discharge is made below the waterline, taking into account the location of the seawater intakes; and

(e) the discharge is made at a distance of not less than 12 nautical miles from the nearest land and in a depth of water of not less than 25 metres.

(4) The discharge into the sea of substances in Category D as defined in Regulation 3(1)(d) of this Annex, or of those provisionally assessed as such, or ballast water, tank washings, or other residues or mixtures containing such substances shall be prohibited except when all the following conditions are satisfied:

(a) the ship is proceeding en route at a speed of at least 7 knots in the case of self-propelled ships or at least 4 knots in the case of ships which are not self-propelled;

(b) such mixtures are of a concentration not greater than one part of the substance in ten parts of water; and

(c) the discharge is made at a distance of not less than 12 nautical miles from the nearest land.

(5) Ventilation procedures approved by the Administration may be used to remove cargo residues from a tank. Such procedures shall be based upon standards developed by the Organization. If subsequent washing of the tank is necessary, the discharge into the sea of the resulting tank washings shall be made in accordance with paragraph (1), (2), (3) or (4) of this Regulation, whichever is applicable.

(6) The discharge into the sea of substances which have not been categorized, provisionally assessed, or evaluated as referred to in Regulation 4(1) of this Annex, or of ballast water, tank washings, or other residues or mixtures containing such substances shall be prohibited.

Categories A, B and C Substances within Special Areas

Subject to the provisions of Regulation 6 of this Annex,

(7) The discharge into the sea of substances in Category A as defined in Regulation 3(1)(a) of this Annex or of those provisionally assessed as such, or ballast water, tank washings, or other residues or mixtures containing such substances shall be prohibited. If tanks containing such substances or mixtures are to be washed the resulting residues shall be discharged to a reception facility which the States bordering the special area shall provide in accordance with Regulation 7 of this Annex, until the concentration of the substance in the effluent to such facility is at or below the residual concentration prescribed for that substance in column IV of Appendix II to this Annex and until the tank is empty. Provided that the residue then remaining in the tank is subsequently diluted by the addition of a volume of water of not less than 5 per cent of the total volume of the tank, it may be discharged into the sea when all the following conditions are also satisfied:

(a) the ship is proceeding en route at a speed of at least 7 knots in the case of self-propelled ships or at least 4 knots in the case of ships which are not self-propelled;

(b) the discharge is made below the waterline, taking into account the location of the seawater intakes; and

(c) the discharge is made at a distance of not less than 12 nautical miles from the nearest land and in a depth of water of not less than 25 metres.

(8) The discharge into the sea of substances in Category B as defined in Regulation 3(1)(b) of this Annex or of those provisionally assessed as such, or ballast water, tank washings, or other residues or mixtures containing such substances shall be prohibited except when all the following conditions are satisfied:

(a) the tank has been washed after unloading with a volume of water of not less than 0.5 per cent of the total volume of the tank, and the resulting residues have been discharged to a reception facility until the tank is empty;

(b) the ship is proceeding en route at a speed of at least 7 knots in the case of self-propelled ships or at least 4 knots in the case of ships which are not self-propelled;

(c) the procedures and arrangements for discharge and washings are approved by the Administration. Such procedures and arrangements shall be based upon standards developed by the Organization and shall ensure that the concentration and rate of discharge of the effluent is such that the concentration of the substance in the wake astern of the ship does not exceed 1 part per million;

(d) the discharge is made below the waterline, taking into account the location of the seawater intakes; and

(e) the discharge is made at a distance of not less than 12 nautical miles from the nearest land and in a depth of water of not less than 25 metres.

(9) The discharge into the sea of substances in Category C as defined in Regulation 3(1)(c) of this Annex or of those provisionally assessed as such, or ballast water, tank washings, or other residues or mixtures containing such substances shall be prohibited except when all the following conditions are satisfied:

(a) the ship is proceeding en route at a speed of at least 7 knots in the case of self-propelled ships or at least 4 knots in the case of ships which are not self-propelled;

(b) the procedures and arrangements for discharge are approved by the Administration. Such procedures and arrangements shall be based upon standards developed by the Organization and shall ensure that the concentration and rate of discharge of the effluent is such that the concentration of the substance in the wake astern of the ship does not exceed 1 part per million;

(c) the maximum quantity of cargo discharged from each tank and its associated piping system does not exceed the maximum quantity approved in accordance with the procedures referred to in sub-paragraph (b) of this paragraph which shall in no case exceed the greater of 1 cubic metre or 1/3,000 of the tank capacity in cubic metres;

(d) the discharge is made below the waterline, taking into account the location of the seawater intakes; and

(e) the discharge is made at a distance of not less than 12 nautical miles from the nearest land and in a depth of water of not less than 25 metres.

(10) Ventilation procedures approved by the Administration may be used to remove cargo residues from a tank. Such procedures shall be based upon standards developed by the Organization. If subsequent washing of the tank is necessary, the discharge into the sea of the resulting tank washings shall be made in accordance with paragraph (7), (8) or (9) of this Regulation, whichever is applicable.

(11) The discharge into the sea of substances which have not been categorized, provisionally assessed or evaluated as referred to in Regulation 4(1) of this Annex, or of ballast water, tank washings, or other residues or mixtures containing such substances shall be prohibited.

(12) Nothing in this Regulation shall prohibit a ship from retaining on board the residues from a Category B or C cargo and discharging such residues into the sea outside a special area in accordance with paragraph (2) or (3) of this Regulation, respectively.

(13) (a) The Governments of Parties to the Convention, the coastlines of which border on any given special area, shall collectively agree and establish a date by which time the requirement of Regulation 7(1) of this Annex will be fulfilled and from which the requirements of paragraphs (7), (8), (9) and (10) of this Regulation in respect of that area shall take effect and notify the Organization of the date so established at least six months in advance of that date. The Organization shall then promptly notify all Parties of that date.

(b) If the date of entry into force of the present Convention is earlier than the date established in accordance with sub-paragraph (a) of this paragraph, the requirements of paragraphs (1), (2) and (3) of this Regulation shall apply during the interim period.

Regulation 6

Exceptions

Regulation 5 of this Annex shall not apply to:

(a) the discharge into the sea of noxious liquid substances or mixtures containing such substances necessary for the purpose of securing the safety of a ship or saving life at sea; or

(b) the discharge into the sea of noxious liquid substances or mixtures containing such substances resulting from damage to a ship or its equipment:

   (i) provided that all reasonable precautions have been taken after the occurrence of the damage or discovery of the discharge for the purpose of preventing or minimizing the discharge; and

   (ii) except if the owner or the Master acted either with intent to cause damage, or recklessly and with knowledge that damage would probably result; or

(c) the discharge into the sea of noxious liquid substances or mixtures containing such substances, approved by the Administration, when being used for the purpose of combating specific pollution incidents in order to minimize the damage from pollution. Any such discharge shall be subject to the approval of any Government in whose jurisdiction it is contemplated the discharge will occur.

Regulation 7

Reception Facilities

(1) The Government of each Party to the Convention undertakes to ensure the provision of reception facilities according to the needs of ships using its ports, terminals or repair ports as follows:

(a) cargo loading and unloading ports and terminals shall have facilities adequate for reception without undue delay to ships of such residues and mixtures containing noxious liquid substances as would remain for disposal from ships carrying them as a consequence of the application of this Annex; and

(b) ship repair ports undertaking repairs to chemical tankers shall have facilities adequate for the reception of residues and mixtures containing noxious liquid substances.

(2) The Government of each Party shall determine the types of facilities provided for the purpose of paragraph (1) of this Regulation at each cargo loading and unloading port, terminal and ship repair port in its territories and notify the Organization thereof.

(3) Each Party shall notify the Organization, for transmission to the Parties concerned, of any case where facilities required under paragraph (1) of this Regulation are alleged to be inadequate.

Regulation 8

Measures of Control

(1) The Government of each Party to the Convention shall appoint or authorize surveyors for the purpose of implementing this Regulation.

Category A Substances in all Areas

(2) (a) If a tank is partially unloaded or unloaded but not cleaned, an appropriate entry shall be made in the Cargo Record Book.

(b) Until that tank is cleaned every subsequent pumping or transfer operation carried out in connexion with that tank shall also be entered in the Cargo Record Book.

(3) If the tank is to be washed:

(a) the effluent from the tank washing operation shall be discharged from the ship to a reception facility at least until the concentration of the substance in the discharge, as indicated by analyses of samples of the effluent taken by the surveyor, has fallen to the residual concentration specified for that substance in Appendix II to this Annex. When the required residual concentration has been achieved, remaining tank washings shall continue to be discharged to the reception facility until the tank is empty. Appropriate entries of these operations shall be made in the Cargo Record Book and certified by the surveyor; and

(b) after diluting the residue then remaining in the tank with at least 5 per cent of the tank capacity of water, this mixture may be discharged into the sea in accordance with the provisions of sub-paragraphs (1)(a), (b) and (c) or 7(a), (b) and (c), whichever is applicable, of Regulation 5 of this Annex. Appropriate entries of these operations shall be made in the Cargo Record Book.

(4) Where the Government of the receiving Party is satisfied that it is impracticable to measure the concentration of the substance in the effluent without causing undue delay to the ship, that Party may accept an alternative procedure as being equivalent to sub-paragraph (3)(a) provided that:

(a) a precleaning procedure for that tank and that substance, based on standards developed by the Organization, is approved by the Administration and that Party is satisfied that such procedure will fulfil the requirements of paragraph (1) or (7), whichever is applicable, of Regulation 5 of this Annex with respect to the attainment of the prescribed residual concentrations;

(b) a surveyor duly authorized by that Party shall certify in the Cargo Record Book that:

(i) the tank, its pump and piping system have been emptied, and that the quantity of cargo remaining in the tank is at or below the quantity on which the approved precleaning procedure referred to in sub-paragraph (ii) of this paragraph has been based;

(ii) precleaning has been carried out in accordance with the precleaning procedure approved by the Administration for that tank and that substance; and

(iii) the tank washings resulting from such precleaning have been discharged to a reception facility and the tank is empty;

(c) the discharge into the sea of any remaining residues shall be in accordance with the provisions of paragraph (3)(b) of this Regulation and an appropriate entry is made in the Cargo Record Book.

Category B Substances outside Special Areas and Category C Substances in all Areas

(5) Subject to such surveillance and approval by the authorized or appointed surveyor as may be deemed necessary by the Government of the Party, the Master of a ship shall, with respect to a Category B substance outside special areas or a Category C substance in all areas, ensure compliance with the following:

(a) If a tank is partially unloaded or unloaded but not cleaned, an appropriate entry shall be made in the Cargo Record Book.

(b) If the tank is to be cleaned at sea:

(i) the cargo piping system serving that tank shall be drained and an appropriate entry made in the Cargo Record Book;

(ii) the quantity of substance remaining in the tank shall not exceed the maximum quantity which may be discharged into the sea for that substance under Regulation 5(2)(c) of this Annex outside special areas in the case of Category B substances, or under Regulations 5(3)(c) and 5(9)(c) outside and within special areas respectively in the case of Category C substances. An appropriate entry shall be made in the Cargo Record Book;

(iii) where it is intended to discharge the quantity of substance remaining into the sea the approved procedures shall be complied with, and the necessary dilution of the substance satisfactory for such a discharge shall be achieved. An appropriate entry shall be made in the Cargo Record Book; or

(iv) where the tank washings are not discharged into the sea, if any internal transfer of tank washings takes place from that tank an appropriate entry shall be made in the Cargo Record Book; and

(v) any subsequent discharge into the sea of such tank washings shall be made in accordance with the requirements of Regulation 5 of this Annex for the appropriate area and Category of substance involved.

(c) If the tank is to be cleaned in port:

(i) the tank washings shall be discharged to a reception facility and an appropriate entry shall be made in the Cargo Record Book; or

(ii) the tank washings shall be retained on board the ship and an appropriate entry shall be made in the Cargo Record Book indicating the location and disposition of the tank washings.

(d) If after unloading a Category C substance within a special area, any residues or tank washings are to be retained on board until the ship is outside the special area, the Master shall so indicate by an appropriate entry in the Cargo Record Book and in this case the procedures set out in Regulation 5(3) of this Annex shall be applicable.

Category B Substances within Special Areas

(6) Subject to such surveillance and approval by the authorized or appointed surveyor as may be deemed necessary by the Government of the Party, the Master of a ship shall, with respect to a Category B substance within a special area, ensure compliance with the following:

   (a) If a tank is partially unloaded or unloaded but not cleaned, an appropriate entry shall be made in the Cargo Record Book.

   (b) Until that tank is cleaned every subsequent pumping or transfer operation carried out in connexion with that tank shall also be entered in the Cargo Record Book.

   (c) If the tank is to be washed, the effluent from the tank washing operation, which shall contain a volume of water not less than 0.5 per cent of the total volume of the tank, shall be discharged from the ship to a reception facility until the tank, its pump and piping system are empty. An appropriate entry shall be made in the Cargo Record Book.

   (d) If the tank is to be further cleaned and emptied at sea, the Master shall:

   (i) ensure that the approved procedures referred to in Regulation 5(8)(c) of this Annex are complied with and that the appropriate entries are made in the Cargo Record Book; and

   (ii) ensure that any discharge into the sea is made in accordance with the requirements of Regulation 5(8) of this Annex and an appropriate entry is made in the Cargo Record Book.

   (e) If after unloading a Category B substance within a special area, any residues or tank washings are to be retained on board until the ship is outside the special area, the Master shall so indicate by an appropriate entry in the Cargo Record Book and in this case the procedures set out in Regulation 5(2) of this Annex shall be applicable.

Category D Substances in all Areas

(7) The Master of a ship shall, with respect to a Category D substance, ensure compliance with the following:

   (a) If a tank is partially unloaded or unloaded but not cleaned, an appropriate entry shall be made in the Cargo Record Book.

   (b) If the tank is to be cleaned at sea:

   (i) the cargo piping system serving that tank shall be drained and an appropriate entry made in the Cargo Record Book;

   (ii) where it is intended to discharge the quantity of substance remaining into the sea, the necessary dilution of the substance satisfactory for such a discharge shall be achieved. An appropriate entry shall be made in the Cargo Record Book; or

   (iii) where the tank washings are not discharged into the sea, if any internal transfer of tank washings takes place from that tank an appropriate entry shall be made in the Cargo Record Book; and

(iv) any subsequent discharge into the sea of such tank washings shall be made in accordance with the requirements of Regulation 5(4) of this Annex.

(c) If the tank is to be cleaned in port:

(i) the tank washings shall be discharged to a reception facility and an appropriate entry shall be made in the Cargo Record Book; or

(ii) the tank washings shall be retained on board the ship and an appropriate entry shall be made in the Cargo Record Book indicating the location and disposition of the tank washings.

Discharge from a Slop Tank

(8) Any residues retained on board in a slop tank, including those from pump room bilges, which contain a Category A substance, or within a special area either a Category A or a Category B substance, shall be discharged to a reception facility in accordance with the provisions of Regulation 5(1), (7) or (8) of this Annex, whichever is applicable. An appropriate entry shall be made in the Cargo Record Book.

(9) Any residues retained on board in a slop tank, including those from pump room bilges, which contain a quantity of a Category B substance outside a special area or a Category C substance in all areas in excess of the aggregate of the maximum quantities specified in Regulation 5(2)(c), (3)(c) or (9)(c) of this Annex, whichever is applicable, shall be discharged to a reception facility. An appropriate entry shall be made in the Cargo Record Book.

Regulation 9

Cargo Record Book

(1) Every ship to which this Annex applies shall be provided with a Cargo Record Book, whether as part of the ship's official log book or otherwise, in the form specified in Appendix IV to this Annex.

(2) The Cargo Record Book shall be completed, on a tank-to-tank basis, whenever any of the following operations with respect to a noxious liquid substance take place in the ship:

(i) loading of cargo;

(ii) unloading of cargo;

(iii) transfer of cargo;

(iv) transfer of cargo, cargo residues or mixtures containing cargo to a slop tank;

(v) cleaning of cargo tanks;

(vi) transfer from slop tanks;

(vii) ballasting of cargo tanks;

(viii) transfer of dirty ballast water;

(ix) discharge into the sea in accordance with Regulation 5 of this Annex.

(3) In the event of any discharge of the kind referred to in Article 7 of the present Convention and Regulation 6 of this Annex of any noxious liquid substance or mixture containing such substance, whether intentional or accidental, an entry shall be made in the Cargo Record Book stating the circumstances of, and the reason for, the discharge.

(4) When a surveyor appointed or authorized by the Government of the Party to the Convention to supervise any operations under this Annex has inspected a ship, then that surveyor shall make an appropriate entry in the Cargo Record Book.

(5) Each operation referred to in paragraphs (2) and (3) of this Regulation shall be fully recorded without delay in the Cargo Record Book so that all the entries in the Book appropriate to that operation are completed. Each entry shall be signed by the officer or officers in charge of the operation concerned and, when the ship is manned, each page shall be signed by the Master of the ship. The entries in the Cargo Record Book shall be in an official language of the State whose flag the ship is entitled to fly, and, for ships holding an International Pollution Prevention Certificate for the Carriage of Noxious Liquid Substances in Bulk (1973) in English or French. The entries in an official national language of the State whose flag the ship is entitled to fly shall prevail in case of a dispute or discrepancy.

(6) The Cargo Record Book shall be kept in such a place as to be readily available for inspection and, except in the case of unmanned ships under tow, shall be kept on board the ship. It shall be retained for a period of two years after the last entry has been made.

(7) The competent authority of the Government of a Party may inspect the Cargo Record Book on board any ship to which this Annex applies while the ship is in its port, and may make a copy of any entry in that book and may require the Master of the ship to certify that the copy is a true copy of such entry. Any copy so made which has been certified by the Master of the ship as a true copy of an entry in the ship's Cargo Record Book shall be made admissible in any judicial proceedings as evidence of the facts stated in the entry. The inspection of a Cargo Record Book and the taking of a certified copy by the competent authority under this paragraph shall be performed as expeditiously as possible without causing the ship to be unduly delayed.

Regulation 10

Surveys

(1) Ships which are subject to the provisions of this Annex and which carry noxious liquid substances in bulk shall be surveyed as follows:

    (a) An initial survey before a ship is put into service or before the certificate required by Regulation 11 of this Annex is issued for the first time, which shall include a complete inspection of its structure, equipment, fittings, arrangements and material in so far as the ship is covered by this Annex. The survey shall be such as to ensure full compliance with the applicable requirements of this Annex.

(b) Periodical surveys at intervals specified by the Administration which shall not exceed five years and which shall be such as to ensure that the structure, equipment, fittings, arrangements and material fully comply with the applicable requirements of this Annex. However, where the duration of the International Pollution Prevention Certificate for the Carriage of Noxious Liquid Substances in Bulk (1973) is extended as specified in Regulation 12(2) or (4) of this Annex, the interval of the periodical survey may be extended correspondingly.

(c) Intermediate surveys at intervals specified by the Administration which shall not exceed thirty months and which shall be such as to ensure that the equipment and associated pump and piping systems, fully comply with the applicable requirements of this Annex and are in good working order. The survey shall be endorsed on the International Pollution Prevention Certificate for the Carriage of Noxious Liquid Substances in Bulk (1973) issued under Regulation 11 of this Annex.

(2) Surveys of a ship with respect to the enforcement of the provisions of this Annex shall be carried out by officers of the Administration. The Administration may, however, entrust the surveys either to surveyors nominated for the purpose or to organizations recognized by it. In every case the Administration concerned shall fully guarantee the completeness and efficiency of the surveys.

(3) After any survey of a ship under this Regulation has been completed, no significant change shall be made in the structure, equipment, fittings, arrangements or material, covered by the survey without the sanction of the Administration, except the direct replacement of such equipment and fittings for the purpose of repair or maintenance.

Regulation 11

Issue of Certificate

(1) An International Pollution Prevention Certificate for the Carriage of Noxious Liquid Substances in Bulk (1973) shall be issued to any ship carrying noxious liquid substances which is engaged in voyages to ports or offshore terminals under the jurisdiction of other Parties to the Convention after survey of such ship in accordance with the provisions of Regulation 10 of this Annex.

(2) Such Certificate shall be issued either by the Administration or by a person or organization duly authorized by it. In every case the Administration shall assume full responsibility for the Certificate.

(3) (a) The Government of a Party may, at the request of the Administration, cause a ship to be surveyed and if satisfied that the provisions of this Annex are complied with shall issue or authorize the issue of a Certificate to the ship in accordance with this Annex.

(b) A copy of the Certificate and a copy of the survey report shall be transmitted as soon as possible to the requesting Administration.

(c) A Certificate so issued shall contain a statement to the effect that it has been issued at the request of the Administration and shall have the same force and receive the same recognition as a certificate issued under paragraph (1) of this Regulation.

(d) No International Pollution Prevention Certificate for the Carriage of Noxious Liquid Substances in Bulk (1973) shall be issued to any ship which is entitled to fly the flag of a State which is not a Party.

(4) The Certificate shall be drawn up in an official language of the issuing country in a form corresponding to the model given in Appendix V to this Annex. If the language used is neither English nor French, the text shall include a translation into one of these languages.

Regulation 12

Duration of Certificate

(1) An International Pollution Prevention Certificate for the Carriage of Noxious Liquid Substances in Bulk (1973) shall be issued for a period specified by the Administration, which shall not exceed five years from the date of issue, except as provided in paragraphs (2) and (4) of this Regulation.

(2) If a ship at the time when the Certificate expires is not in a port or offshore terminal under the jurisdiction of the Party to the Convention whose flag the ship is entitled to fly, the Certificate may be extended by the Administration, but such extension shall be granted only for the purpose of allowing the ship to complete its voyage to the State whose flag the ship is entitled to fly or in which it is to be surveyed and then only in cases where it appears proper and reasonable to do so.

(3) No Certificate shall be thus extended for a period longer than five months and a ship to which such extension is granted shall not on its arrival in the State whose flag it is entitled to fly or the port in which it is to be surveyed, be entitled by virtue of such extension to leave that port or State without having obtained a new Certificate.

(4) A Certificate which has not been extended under the provisions of paragraph (2) of this Regulation may be extended by the Administration for a period of grace of up to one month from the date of expiry stated on it.

(5) A Certificate shall cease to be valid if significant alterations have taken place in the structure, equipment, fittings, arrangements and material required by this Annex without the sanction of the Administration, except the direct replacement of such equipment or fitting for the purpose of repair or maintenance or if intermediate surveys as specified by the Administration under Regulation 10(1)(c) of this Annex are not carried out.

(6) A Certificate issued to a ship shall cease to be valid upon transfer of such a ship to the flag of another State, except as provided in paragraph (7) of this Regulation.

(7) Upon transfer of a ship to the flag of another Party, the Certificate shall remain in force for a period not exceeding five months provided that it would not have expired before the end of that period, or until the Administration issues a replacement Certificate, whichever is earlier. As soon as possible after the transfer has taken place the Government of the Party whose flag the ship was formerly entitled to fly shall transmit to the Administration a copy of the Certificate carried by the ship before the transfer and, if available, a copy of the relevant survey report.

Regulation 13

Requirements for Minimizing accidental Pollution

(1) The design, construction, equipment and operation of ships carrying noxious liquid substances in bulk which are subject to the provisions of this Annex shall be such as to minimize the uncontrolled discharge into the sea of such substances.

(2) Pursuant to the provisions of paragraph (1) of this Regulation, the Government of each Party shall issue, or cause to be issued, detailed requirements on the design, construction, equipment and operation of such ships.

(3) In respect of chemical tankers, the requirements referred to in paragraph (2) of this Regulation shall contain at least all the provisions given in the Code for the Construction and Equipment of Ships carrying Dangerous Chemicals in Bulk adopted by the Assembly of the Organization in Resolution A.212(VII) and as may be amended by the Organization, provided that the amendments to that Code are adopted and brought into force in accordance with the provisions of Article 16 of the present Convention for amendment procedures to an Appendix to an Annex.

Appendix I

GUIDELINES FOR THE CATEGORIZATION OF
NOXIOUS LIQUID SUBSTANCES

Category A Substances which are bioaccumulated and liable to produce a hazard to aquatic life or human health; or which are highly toxic to aquatic life (as expressed by a Hazard Rating 4, defined by a TLm less than 1 ppm); and additionally certain substances which are moderately toxic to aquatic life (as expressed by a Hazard Rating 3, defined by a TLm of 1 or more, but less than 10 ppm) when particular weight is given to additional factors in the hazard profile or to special characteristics of the substance.

Category B Substances which are bioaccumulated with a short retention of the order of one week or less; or which are liable to produce tainting of the sea food; or which are moderately toxic to aquatic life (as expressed by a Hazard Rating 3, defined by a TLm of 1 ppm or more, but less than 10 ppm); and additionally certain substances which are slightly toxic to aquatic life (as expressed by a Hazard Rating 2, defined by a TLm of 10 ppm or more, but less than 100 ppm) when particular weight is given to additional factors in the hazard profile or to special characteristics of the substance.

Category C Substances which are slightly toxic to aquatic life (as expressed by a Hazard Rating 2, defined by a TLm of 10 or more, but less than 100 ppm); and additionally certain substances which are practically non-toxic to aquatic life (as expressed by a Hazard Rating 1, defined by a TLm of 100 ppm or more, but less than 1,000 ppm) when particular weight is given to additional factors in the hazard profile or to special characteristics of the substance.

Category D Substances which are practically non-toxic to aquatic life, (as expressed by a Hazard Rating 1, defined by a TLm of 100 ppm or more, but less than 1,000 ppm); or causing deposits blanketing the seafloor with a high biochemical oxygen demand (BOD); or highly hazardous to human health, with an LD_{50} of less than 5 mg/kg; or produce moderate reduction of amenities because of persistency, smell or poisonous or irritant characteristics, possibly interfering with use of beaches; or moderately hazardous to human health, with an LD_{50} of 5 mg/kg or more, but less than 50 mg/kg and produce slight reduction of amenities.

Other Liquid Substances (for the purposes of Regulation 4 of this Annex)
Substances other than those categorized in Categories A, B, C and D above.

Appendix II

LIST OF NOXIOUS LIQUID SUBSTANCES CARRIED IN BULK

Substance	UN Number	Pollution Category for operational discharge (Regulation 3 of Annex II)	Residual concentration (per cent by weight) (Regulation 5(1) of Annex II) Outside special areas	Residual concentration (per cent by weight) (Regulation 5(7) of Annex II) Within special areas
I		II	III	IV
Acetaldehyde	1089	C		
Acetic acid	1842	C		
Acetic anhydride	1715	C		
Acetone	1090	D		
Acetone cyanohydrin	1541	A	0.1	0.05
Acetyl chloride	1717	C		
Acrolein	1092	A	0.1	0.05
Acrylic acid*	-	C		
Acrylonitrile	1093	B		
Adiponitrile	-	D		
Alkylbenzene sulfonate (straight chain)	-	C		
(branched chain)		B		
Allyl alcohol	1098	B		
Allyl chloride	1100	C		
Alum (15% solution)	-	D		
Aminoethylethanolamine (Hydroxyethyl-ethylene-diamine)*	-	D		
Ammonia (28% aqueous)	1005	B		
iso-Amyl acetate	1104	C		
n-Amyl acetate	1104	C		
n-Amyl alcohol	-	D		
Aniline	1547	C		

* Asterisk indicates that the substance has been provisionally included in this list and that further data are necessary in order to complete the evaluation of its environmental hazards, particularly in relation to living resources.

Substance	I	II	III	IV
Benzene	1114	C		
Benzyl alcohol	-	D		
Benzyl chloride	1738	B		
n-Butyl acetate	1123	D		
sec-Butyl acetate	1124	D		
n-Butyl acrylate	-	D		
Butyl butyrate*	-	B		
Butylene glycol(s)	-	D		
Butyl methacrylate	-	D		
n-Butyraldehyde	1129	B		
Butyric acid	-	B		
Calcium hydroxide (solution)	-	D		
Camphor oil	1130	B		
Carbon disulphide	1131	A	0.01	0.005
Carbon tetrachloride	1846	B		
Caustic potash (Potassium hydroxide)	1814	C		
Chloroacetic acid	1750	C		
Chloroform	1888	B		
Chlorohydrins (crude)*	-	D		
Chloroprene*	1991	C		
Chlorosulphonic acid	1754	C		
para-Chlorotoluene	-	B		
Citric acid (10%-25%)	-	D		
Creosote	1334	A	0.1	0.05
Cresols	2076	A	0.1	0.05
Cresylic acid	2022	A	0.1	0.05
Crotonaldehyde	1143	B		
Cumene	1918	C		
Cyclohexane	1145	C		
Cyclohexanol	-	D		
Cyclohexanone	1915	D		
Cyclohexylamine*	-	D		
para-Cymene (Isopropyltoluene)*	2046	D		
Decahydronaphthalene	1147	D		
Decane*	-	D		

---

* Asterisk indicates that the substance has been provisionally included in this list and that further data are necessary in order to complete the evaluation of its environmental hazards, particularly in relation to living resources.

Substance	I	II	III	IV
Diacetone alcohol*	1148	D		
Dibenzyl ether*	-	C		
Dichlorobenzenes	1591	A	0.1	0.05
Dichloroethyl ether	1916	B		
Dichloropropene - Dichloropropane mixture (D.D. Soil fumigant)	2047	B		
Diethylamine	1154	C		
Diethylbenzene (mixed isomers)	2049	C		
Diethyl ether	1155	D		
Diethylenetriamine*	2079	C		
Diethylene glycol monoethyl ether	-	C		
Diethylketone (3-Pentanone)	1156	D		
Diisobutylene*	2050	D		
Diisobutyl ketone	1157	D		
Diisopropanolamine	-	C		
Diisopropylamine	1158	C		
Diisopropyl ether*	1159	D		
Dimethylamine (40% aqueous)	1160	C		
Dimethylethanolamine (2-Dimethylamino-ethanol)*	2051	C		
Dimethylformamide	-	D		
1,4-Dioxane*	1165	C		
Diphenyl/Diphenyloxide, mixtures*	-	D		
Dodecylbenzene	-	C		
Epichlorohydrin	2023	B		
2-Ethoxyethyl acetate*	1172	D		
Ethyl acetate	1173	D		
Ethyl acrylate	1917	D		
Ethyl amyl ketone*	-	C		
Ethylbenzene	1175	C		
Ethyl cyclohexane	-	D		

---

* Asterisk indicates that the substance has been provisionally included in this list and that further data are necessary in order to complete the evaluation of its environmental hazards, particularly in relation to living resources.

Substance	I	II	III	IV
Ethylene chlorohydrin (2-Chloro-ethanol)	1135	D		
Ethylene cyanohydrin*	-	D		
Ethylenediamine	1604	C		
Ethylene dibromide	1605	B		
Ethylene dichloride	1184	B		
Ethylene glycol monoethyl ether (Methyl cellosolve)	1171	D		
2-Ethylhexyl acrylate*	-	D		
2-Ethylhexyl alcohol	-	C		
Ethyl lactate*	1192	D		
2-Ethyl 3-propyl-acrolein*	-	B		
Formaldehyde (37-50% solution)	1198	C		
Formic acid	1779	D		
Furfuryl alcohol	-	C		
Heptanoic acid*	-	D		
Hexamethylenediamine*	1783	C		
Hydrochloric acid	1789	D		
Hydrofluoric acid (40% aqueous)	1790	B		
Hydrogen peroxide (greater than 60%)	2015	C		
Isobutyl acrylate	-	D		
Isobutyl alcohol	1212	D		
Isobutyl methacrylate	-	D		
Isobutyraldehyde	2045	C		
Isooctane*	-	D		
Isopentane	-	D		
Isophorone	-	D		
Isopropylamine	1221	C		
Isopropyl cyclohexane	-	D		
Isoprene	1218	D		
Lactic acid	-	D		
Mesityl oxide*	1229	C		
Methyl acetate	1231	D		
Methyl acrylate	1919	C		
Methylamyl alcohol	-	D		

* Asterisk indicates that the substance has been provisionally included in this list and that further data are necessary in order to complete the evaluation of its environmental hazards, particularly in relation to living resources.

Substance	I	II	III	IV
Methylene chloride	1593	B		
2-Methyl-5-Ethyl-pyridine*	-	B		
Methyl methacrylate	1247	D		
2-Methylpentene*	-	D		
alpha-Methylstyrene*	-	D		
Monochlorobenzene	1134	B		
Monoethanolamine	-	D		
Monoisopropanolamine	-	C		
Monomethyl ethanolamine	-	C		
Mononitrobenzene	-	C		
Monoisopropylamine	-	C		
Morpholine*	2054	C		
Naphthalene (molten)	1334	A	0.1	0.05
Naphthenic acids*	-	A	0.1	0.05
Nitric acid (90%)	2031/2032	C		
2-Nitropropane	-	D		
ortho-Nitrotoluene	1664	C		
Nonyl alcohol*	-	C		
Nonylphenol	-	C		
n-Octanol	-	C		
Oleum	1831	C		
Oxalic acid (10–25%)	-	D		
Pentachloroethane	1669	B		
n-Pentane	1265	C		
Perchloroethylene (Tetrachloroethylene)	1897	B		
Phenol	1671	B		
Phosphoric acid	1805	D		
Phosphorus (elemental)	1338	A	0.01	0.005
Phthalic anhydride (molten)	-	C		
beta-Propiolactone*	-	B		
Propionaldehyde	1275	D		
Propionic acid	1848	D		
Propionic anhydride	-	D		
n-Propyl acetate*	1276	C		

* Asterisk indicates that the substance has been provisionally included in this list and that further data are necessary in order to complete the evaluation of its environmental hazards, particularly in relation to living resources.

Substance	I	II	III	IV
n-Propyl alcohol	1274	D		
n-Propylamine	1277	C		
Pyridine	1282	B		
Silicon tetrachloride	1818	D		
Sodium bichromate (solution)	-	C		
Sodium hydroxide	1824	C		
Sodium pentachlorophenate (solution)	-	A	0.1	0.05
Styrene monomer	2055	C		
Sulphuric acid	1830/ 1831/ 1832	C		
Tallow	-	D		
Tetraethyl lead	1649	A	0.1	0.05
Tetrahydrofuran	2056	D		
Tetrahydronaphthalene	1540	C		
Tetramethylbenzene	-	D		
Tetramethyl lead	1649	A	0.1	0.05
Titanium tetrachloride	1838	D		
Toluene	1294	C		
Toluene diisocyanate*	2078	B		
Trichloroethane	-	C		
Trichloroethylene	1710	B		
Triethanolamine	-	D		
Triethylamine	1296	C		
Trimethylbenzene*	-	C		
Tritolyl phosphate (Tricresyl phosphate)*	-	B		
Turpentine (wood)	1299	B		
Vinyl acetate	1301	C		
Vinylidene chloride*	1303	B		
Xylenes (mixed isomers)	1307	C		

* Asterisk indicates that the substance has been provisionally included in this list and that further data are necessary in order to complete the evaluation of its environmental hazards, particularly in relation to living resources.

Appendix III

LIST OF OTHER LIQUID SUBSTANCES CARRIED IN BULK

Acetonitrile (Methyl cyanide)
tert-Amyl alcohol
n-Butyl alcohol
Butyrolactone
Calcium chloride (solution)
Castor oil
Citric juices
Coconut oil
Cod liver oil
iso-Decyl alcohol
n-Decyl alcohol
Decyl octyl alcohol
Dibutyl ether
Diethanolamine
Diethylene glycol
Dipentene
Dipropylene glycol
Ethyl alcohol
Ethylene glycol
Fatty alcohols (C_{12}-C_{20})
Glycerine
n-Heptane
Heptene (mixed isomers)
n-Hexane
Ligroin
Methyl alcohol
Methylamyl acetate
Methyl ethyl ketone (2-butanone)
Milk
Molasses

Olive Oil
Polypropylene glycol
iso-Propyl acetate
iso-Propyl alcohol
Propylene glycol
Propylene oxide
Propylene tetramer
Propylene trimer
Sorbitol
Sulphur (liquid)
Tridecanol
Triethylene glycol
Triethylenetetramine
Tripropylene glycol
Water
Wine

Appendix IV

CARGO RECORD BOOK FOR SHIPS CARRYING NOXIOUS LIQUID SUBSTANCES IN BULK

Name of ship..

Cargo carrying capacity of
each tank in cubic metres ....

Voyage from .... to ....

(a) **Loading of cargo**

    1. Date and place of loading

    2. Name and category of cargo(es) loaded

    3. Identity of tank(s) loaded

(b) **Transfer of cargo**

    4. Date of transfer

    5. Identity of tank(s) (i) From
                            (ii) To

    6. Was(were) tank(s) in 5(i) emptied?

    7. If not, quantity remaining

(c) **Unloading of cargo**

    8. Date and place of unloading

    9. Identity of tank(s) unloaded

    10. Was(were) tank(s) emptied?

    11. If not, quantity remaining in tank(s)

    12. Is(are) tank(s) to be cleaned?

    13. Amount transferred to slop tank

    14. Identity of slop tank

(d) **Ballasting of cargo tanks**

    15. Identity of tank(s) ballasted

    16. Date and position of ship at start of ballasting

.... Signature of Master

(e) **Cleaning of cargo tanks**

   Category A substances

   17. Identity of tank(s) cleaned
   18. Date and location of cleaning
   19. Method(s) of cleaning
   20. Location of reception facility used
   21. Concentration of effluent when discharge to reception facility stopped
   22. Quantity remaining in tank
   23. Procedure and amount of water introduced into tank in final cleaning
   24. Location, date of discharge into sea
   25. Procedure and equipment used in discharge into the sea

   Category B, C and D substances

   26. Washing procedure used
   27. Quantity of water used
   28. Date, location of discharge into sea
   29. Procedure and equipment used in discharge into the sea

(f) **Transfer of dirty ballast water**

   30. Identity of tank(s)
   31. Date and position of ship at start of discharge into sea
   32. Date and position of ship at finish of discharge into sea
   33. Ship's speed(s) during discharge
   34. Quantity discharged into sea
   35. Quantity of polluted water transferred to slop tank(s) (identify slop tank(s))
   36. Date and port of discharge to shore reception facilities (if applicable)

.... Signature of Master

(g) **Transfer from slop tank/disposal of residue**
    37. Identity of slop tank(s)
    38. Quantity disposed from each tank
    39. Method of disposal of residue:
        (a) Reception facilities
        (b) Mixed with cargo
        (c) Transferred to another (other) tank(s) (identify tank(s))
        (d) Other method
    40. Date and port of disposal of residue

(h) **Accidental or other exceptional discharge**
    41. Date and time of occurrence
    42. Place or position of ship at time of occurrence
    43. Approximate quantity, name and category of substance
    44. Circumstances of discharge or escape and general remarks.

.... Signature of Master

Appendix V

FORM OF CERTIFICATE

INTERNATIONAL POLLUTION PREVENTION CERTIFICATE FOR THE CARRIAGE OF NOXIOUS LIQUID SUBSTANCES IN BULK (1973)

(*Note:* This Certificate shall be supplemented in the case of a chemical tanker by the certificate required pursuant to the provisions of Regulation 13(3) of Annex II of the Convention)

(Official Seal)

Issued under the provisions of the International Convention for the Prevention of Pollution from Ships, 1973, under the authority of the Government of

....
(full official designation of the country)

by ....
(full official designation of the competent person or organization authorized under the provisions of the International Convention for the Prevention of Pollution from Ships, 1973)

Name of Ship	Distinctive Number or Letter	Port of Registry	Gross Tonnage

THIS IS TO CERTIFY:

1. That the ship has been surveyed in accordance with the provisions of Regulation 10 of Annex II of the Convention.

2. That the survey showed that the design, construction and equipment of the ship are such as to minimize the uncontrolled discharge into the sea of noxious liquid substances.

3. That the following arrangements and procedures have been approved by the Administration in connexion with the implementation of Regulation 5 of Annex II of the Convention:

....

(Continued on the annexed signed and dated sheet(s))

....

This certificate is valid, until ....,
subject to intermediate survey(s) at intervals of ....

Issued at...
(place of issue of Certificate)

.....................19..   ....
                                *(Signature of duly authorized official
                                issuing the Certificate)*

(Seal or stamp of the issuing Authority, as appropriate)

Intermediate surveys

This is to certify that at an intermediate survey required by Regulation 10(1)(c) of Annex II of the Convention, this ship and the condition thereof are found to comply with the relevant provisions of the Convention.

                                Signed............................
                                    (Signature of duly authorized official)

                                Place ....

                                Date ....

(Seal or stamp of the Authority, as appropriate)

                                Signed............................
                                    (Signature of duly authorized official)

                                Place ....

                                Date ....

(Seal or stamp of the Authority, as appropriate)

Under the provisions of Regulation 12(2) and (4) of Annex II of the Convention the validity of this Certificate is extended until

....

                                Signed............................
                                    (Signature of duly authorized official)

                                Place ....

                                Date ....

(Seal or stamp of the Authority, as appropriate)

ANNEX III

REGULATIONS FOR THE PREVENTION OF POLLUTION BY HARMFUL SUBSTANCES CARRIED BY SEA IN PACKAGED FORMS, OR IN FREIGHT CONTAINERS, PORTABLE TANKS OR ROAD AND RAIL TANK WAGONS

Regulation 1

Application

(1) Unless expressly provided otherwise, the Regulations of this Annex apply to all ships carrying harmful substances in packaged forms, or in freight containers, portable tanks or road and rail tank wagons.

(2) Such carriage of harmful substances is prohibited except in accordance with the provisions of this Annex.

(3) To supplement the provisions of this Annex the Government of each Party to the Convention shall issue, or cause to be issued, detailed requirements on packaging, marking and labelling, documentation, stowage, quantity limitations, exceptions and notification, for preventing or minimizing pollution of the marine environment by harmful substances.

(4) For the purpose of this Annex, empty receptacles, freight containers, portable tanks and road and rail tank wagons which have been used previously for the carriage of harmful substances shall themselves be treated as harmful substances unless adequate precautions have been taken to ensure that they contain no residue that is hazardous to the marine environment.

Regulation 2

Packaging

Packagings, freight containers, portable tanks and road and rail tank wagons shall be adequate to minimize the hazard to the marine environment having regard to their specific contents.

Regulation 3

Marking and Labelling

Packages, whether shipped individually or in units or in freight containers, freight containers, portable tanks or road and rail tank wagons containing a harmful substance, shall be durably marked with the correct technical name (trade names shall not be used as the correct technical name), and further marked with a distinctive label or stencil of label, indicating that the contents are harmful. Such identification shall be supplemented where possible by any other means, for example by the use of the United Nations number.

Regulation 4

Documentation

(1) In all documents relating to the carriage of harmful substances by sea where such substances are named, the correct technical name of the substances shall be used (trade names shall not be used).

(2) The shipping documents supplied by the shipper shall include a certificate or declaration that the shipment offered for carriage is properly packed, marked and labelled and in proper condition for carriage to minimize the hazard to the marine environment.

(3) Each ship carrying harmful substances shall have a special list or manifest setting forth the harmful substances on board and the location thereof. A detailed stowage plan which sets out the location of all harmful substances on board may be used in place of such special list or manifest. Copies of such documents shall also be retained on shore by the owner of the ship or his representative until the harmful substances are unloaded.

(4) In a case where the ship carries a special list or manifest or a detailed stowage plan, required for the carriage of dangerous goods by the International Convention for the Safety of Life at Sea in force, the documents required for the purpose of this Annex may be combined with those for dangerous goods. Where documents are combined, a clear distinction shall be made between dangerous goods and other harmful substances.

Regulation 5

Stowage

Harmful substances shall be both properly stowed and secured so as to minimize the hazards to the marine environment without impairing the safety of ship and persons on board.

Regulation 6

Quantity Limitations

Certain harmful substances which are very hazardous to the marine environment may, for sound scientific and technical reasons, need to be prohibited for carriage or be limited as to the quantity which may be carried aboard any one ship. In limiting the quantity due consideration shall be given to size, construction and equipment of the ship as well as the packaging and the inherent nature of the substance.

Regulation 7

Exceptions

(1) Discharge by jettisoning of harmful substances carried in packaged forms, freight containers, portable tanks or road and rail tank wagons shall be prohibited except where necessary for the purpose of securing the safety of the ship or saving life at sea.

(2) Subject to the provisions of the present Convention, appropriate measures based on the physical, chemical and biological properties of harmful substances shall be taken to regulate the washing of leakages overboard provided that compliance with such measures would not impair the safety of the ship and persons on board.

Regulation 8

Notification

With respect to certain harmful substances, as may be designated by the Government of a Party to the Convention, the master or owner of the ship or his representative shall notify the appropriate port authority of the intent to load or unload such substances at least 24 hours prior to such action.

ANNEX IV

REGULATIONS FOR THE PREVENTION OF POLLUTION BY SEWAGE FROM SHIPS

Regulation 1

Definitions

For the purposes of the present Annex:

(1) "New ship" means a ship:

    (a) for which the building contract is placed, or in the absence of a building contract, the keel of which is laid, or which is at a similar stage of construction, on or after the date of entry into force of this Annex; or

    (b) the delivery of which is three years or more after the date of entry into force of this Annex.

(2) "Existing ship" means a ship which is not a new ship.

(3) "Sewage" means:

    (a) drainage and other wastes from any form of toilets, urinals, and WC scuppers;

    (b) drainage from medical premises (dispensary, sick bay, etc.) via wash basins, wash tubs and scuppers located in such premises;

    (c) drainage from spaces containing living animals; or

    (d) other waste waters when mixed with the drainages defined above.

(4) "Holding tank" means a tank used for the collection and storage of sewage.

(5) "Nearest land". The term "from the nearest land" means from the baseline from which the territorial sea of the territory in question is established in accordance with international law except that, for the purposes of the present Convention "from the nearest land" off the north eastern coast of Australia shall mean from a line drawn from a point on the coast of Australia in

    latitude 11° 00′ South, longitude 142°08′ East to a point in latitude 10°35′ South,
    longitude 141° 55′ East – thence to a point latitude 10° 00′ South, longitude 142° 00′ East, thence to a point latitude 9° 10′ South, longitude 143° 52′ East, thence to a point latitude 9° 00′ South, longitude 144° 30′ East, thence to a point latitude 13° 00′ South, longitude 144° 00′ East, thence to a point latitude 15° 00′ South, longitude 146° 00′ East, thence to a point latitude 18° 00′ South, longitude 147° 00′ East, thence to a point latitude 21° 00′ South, longitude 153° 00′ East, thence to a point on the coast of Australia in latitude 24°42′ South, longitude 153°15′ East.

Regulation 2

Application

The provisions of this Annex shall apply to:

(a) (i) new ships of 200 tons gross tonnage and above;

    (ii) new ships of less than 200 tons gross tonnage which are certified to carry more than 10 persons;

    (iii) new ships which do not have a measured gross tonnage and are certified to carry more than 10 persons; and

(b) (i) existing ships of 200 tons gross tonnage and above, 10 years after the date of entry into force of this Annex;

    (ii) existing ships of less than 200 tons gross tonnage which are certified to carry more than 10 persons, 10 years after the date of entry into force of this Annex; and

    (iii) existing ships which do not have a measured gross tonnage and are certified to carry more than 10 persons, 10 years after the date of entry into force of this Annex.

Regulation 3

Surveys

(1) Every ship which is required to comply with the provisions of this Annex and which is engaged in voyages to ports or offshore terminals under the jurisdiction of other Parties to the Convention shall be subject to the surveys specified below:

(a) An initial survey before the ship is put in service or before the Certificate required under Regulation 4 of this Annex is issued for the first time, which shall include a survey of the ship which shall be such as to ensure:

    (i) when the ship is equipped with a sewage treatment plant the plant shall meet operational requirements based on standards and the test methods developed by the Organization;

    (ii) when the ship is fitted with a system to comminute and disinfect the sewage, such a system shall be of a type approved by the Administration;

    (iii) when the ship is equipped with a holding tank the capacity of such tank shall be to the satisfaction of the Administration for the retention of all sewage having regard to the operation of the ship, the number of persons on board and other relevant factors. The holding tank shall have a means to indicate visually the amount of its contents; and

    (iv) that the ship is equipped with a pipeline leading to the exterior convenient for the discharge of sewage to a reception facility and that such a pipeline is fitted with a standard shore connection in compliance with Regulation 11 of this Annex.

This survey shall be such as to ensure that the equipment, fittings, arrangements and material fully comply with the applicable requirements of this Annex.

(b) Periodical surveys at intervals specified by the Administration but not exceeding five years which shall be such as to ensure that the equipment, fittings, arrangements and material fully comply with the applicable requirements of this Annex. However, where the duration of the International Sewage Pollution Prevention Certificate (1973) is extended as specified in Regulation 7(2) or (4) of this Annex, the interval of the periodical survey may be extended correspondingly.

(2) The Administration shall establish appropriate measures for ships which are not subject to the provisions of paragraph (1) of this Regulation in order to ensure that the provisions of this Annex are complied with.

(3) Surveys of the ship as regards enforcement of the provisions of this Annex shall be carried out by officers of the Administration. The Administration may, however, entrust the surveys either to surveyors nominated for the purpose or to organizations recognized by it. In every case the Administration concerned fully guarantees the completeness and efficiency of the surveys.

(4) After any survey of the ship under this Regulation has been completed, no significant change shall be made in the equipment, fittings, arrangements, or material covered by the survey without the approval of the Administration, except the direct replacement of such equipment or fittings.

Regulation 4

Issue of Certificate

(1) An International Sewage Pollution Prevention Certificate (1973) shall be issued, after survey in accordance with the provisions of Regulation 3 of this Annex, to any ship which is engaged in voyages to ports or offshore terminals under the jurisdiction of other Parties to the Convention.

(2) Such Certificate shall be issued either by the Administration or by any persons or organization duly authorized by it. In every case the Administration assumes full responsibility for the Certificate.

Regulation 5

Issue of a Certificate by another Government

(1) The Government of a Party to the Convention may, at the request of the Administration, cause a ship to be surveyed and, if satisfied that the provisions of this Annex are complied with, shall issue or authorize the issue of an International Sewage Pollution Prevention Certificate (1973) to the ship in accordance with this Annex.

(2) A copy of the Certificate and a copy of the survey report shall be transmitted as early as possible to the Administration requesting the survey.

(3) A Certificate so issued shall contain a statement to the effect that it has been issued at the request of the Administration and it shall have the same force and receive the same recognition as the Certificate issued under Regulation 4 of this Annex.

(4) No International Sewage Pollution Prevention Certificate (1973) shall be issued to a ship which is entitled to fly the flag of a State, which is not a Party.

Regulation 6

Form of Certificate

The International Sewage Pollution Prevention Certificate (1973) shall be drawn up in an official language of the issuing country in the form corresponding to the model given in the Appendix to this Annex. If the language used is neither English nor French, the text shall include a translation into one of these languages.

Regulation 7

Duration of Certificate

(1) An International Sewage Pollution Prevention Certificate (1973) shall be issued for a period specified by the Administration, which shall not exceed five years from the date of issue, except as provided in paragraphs (2), (3) and (4) of this Regulation.

(2) If a ship at the time when the Certificate expires is not in a port or offshore terminal under the jurisdiction of the Party to the Convention whose flag the ship is entitled to fly, the Certificate may be extended by the Administration, but such extension shall be granted only for the purpose of allowing the ship to complete its voyage to the State whose flag the ship is entitled to fly or in which it is to be surveyed and then only in cases where it appears proper and reasonable to do so.

(3) No Certificate shall be thus extended for a period longer than five months and a ship to which such extension is granted shall not on its arrival in the State whose flag it is entitled to fly or the port in which it is to be surveyed, be entitled by virtue of such extension to leave that port or State without having obtained a new Certificate.

(4) A Certificate which has not been extended under the provisions of paragraph (2) of this Regulation may be extended by the Administration for a period of grace of up to one month from the date of expiry stated on it.

(5) A Certificate shall cease to be valid if significant alterations have taken place in the equipment, fittings, arrangement or material required without the approval of the Administration, except the direct replacement of such equipment or fittings.

(6) A Certificate issued to a ship shall cease to be valid upon transfer of such a ship to the flag of another State, except as provided in paragraph (7) of this Regulation.

(7) Upon transfer of a ship to the flag of another Party, the Certificate shall remain in force for a period not exceeding five months provided that it would not have expired before the end of that period, or until the Administration issues a replacement Certificate, whichever is earlier. As soon as possible after the transfer has taken place the Government of the Party whose flag the ship was formerly entitled to fly shall transmit to the Administration a copy of the Certificate carried by the ship before the transfer and, if available, a copy of the relevant survey report.

Regulation 8

Discharge of Sewage

(1) Subject to the provisions of Regulation 9 of this Annex, the discharge of sewage into the sea is prohibited, except when:

- (a) the ship is discharging comminuted and disinfected sewage using a system approved by the Administration in accordance with Regulation 3(1)(a) at a distance of more than four nautical miles from the nearest land, or sewage which is not comminuted or disinfected at a distance of more than 12 nautical miles from the nearest land, provided that in any case, the sewage that has been stored in holding tanks shall not be discharged instantaneously but at a moderate rate when the ship is en route and proceeding at not less than 4 knots; the rate of discharge shall be approved by the Administration based upon standards developed by the Organization; or

- (b) the ship has in operation an approved sewage treatment plant which has been certified by the Administration to meet the operational requirements referred to in Regulation 3(1)(a)(i) of this Annex, and
  - (i) the test results of the plant are laid down in the ship's International Sewage Pollution Prevention Certificate (1973);
  - (ii) additionally, the effluent shall not produce visible floating solids in, nor cause discolouration of, the surrounding water; or

- (c) the ship is situated in the waters under the jurisdiction of a State and is discharging sewage in accordance with such less stringent requirements as may be imposed by such State.

(2) When the sewage is mixed with wastes or waste water having different discharge requirements, the more stringent requirements shall apply.

Regulation 9

Exceptions

Regulation 8 of this Annex shall not apply to:

- (a) the discharge of sewage from a ship necessary for the purpose of securing the safety of a ship and those on board or saving life at sea; or

(b) the discharge of sewage resulting from damage to a ship or its equipment if all reasonable precautions have been taken before and after the occurrence of the damage, for the purpose of preventing or minimizing the discharge.

Regulation 10

Reception Facilities

(1) The Government of each Party to the Convention undertakes to ensure the provision of facilities at ports and terminals for the reception of sewage, without causing undue delay to ships, adequate to meet the needs of the ships using them.

(2) The Government of each Party shall notify the Organization for transmission to the Contracting Governments concerned of all cases where the facilities provided under this Regulation are alleged to be inadequate.

Regulation 11

Standard Discharge Connections

To enable pipes of reception facilities to be connected with the ship's discharge pipeline, both lines shall be fitted with a standard discharge connection in accordance with the following table:

STANDARD DIMENSIONS OF FLANGES FOR DISCHARGE CONNECTIONS

Description	Dimension
Outside diameter	210 mm
Inner diameter	According to pipe outside diameter
Bolt circle diameter	170 mm
Slots in flange	4 holes 18 mm in diameter equidistantly placed on a bolt circle of the above diameter, slotted to the flange periphery. The slot width to be 18 mm
Flange thickness	16 mm
Bolts and nuts: quantity and diameter	4, each of 16 mm in diameter and of suitable length
The flange is designed to accept pipes up to a maximum internal diameter of 100 mm and shall be of steel or other equivalent material having a flat face. This flange, together with a suitable gasket, shall be suitable for a service pressure of 6 kg/cm^2.	

For ships having a moulded depth of 5 metres and less, the inner diameter of the discharge connection may be 38 millimetres.

Appendix

FORM OF CERTIFICATE

INTERNATIONAL SEWAGE POLLUTION PREVENTION CERTIFICATE (1973)

Issued under the Provisions of the International Convention for the Prevention of Pollution from Ships, 1973, under the Authority of the Government of

....
(full designation of the country)

by ....
(full designation of the competent person or organization authorized under the provisions of the International Convention for the Prevention of Pollution from Ships, 1973)

Name of Ship	Distinctive Number or Letter	Port of Registry	Gross Tonnage	Number of persons which the ship is certified to carry

New/existing ship*
Date of building contract ....
Date on which keel was laid or ship
was at a similar stage of construction ....
Date of delivery ....

---

* Delete as appropriate

THIS IS TO CERTIFY THAT:

(1) The ship is equipped with a sewage treatment plant/comminuter/holding tank*
and a discharge pipeline in compliance with Regulation 3(1)(a)(i) to (iv) of
Annex IV of the Convention as follows:

   *(a) Description of the sewage treatment plant:
   Type of sewage treatment plant ....
   Name of manufacturer ....
   The sewage treatment plant is certified by the Administration to meet
   the following effluent standards:** ....

   *(b) Description of comminuter:
   Type of comminuter ....
   Name of manufacturer ....
   Standard of sewage after disinfection ....

   *(c) Description of holding tank equipment:
   Total capacity of the holding tank .... m^3
   Location ....

   (d) A pipeline for the discharge of sewage to a reception facility, fitted
   with a standard shore connection.

(2) The ship has been surveyed in accordance with Regulation 3 of Annex IV of
the International Convention for the Prevention of Pollution from Ships,
1973, concerning the prevention of pollution by sewage and the survey
showed that the equipment of the ship and the condition thereof are in all
respects satisfactory and the ship complies with the applicable requirements
of Annex IV of the Convention.

This Certificate is valid until ....

Issued at ....
(place of issue of Certificate)
.... 19 ..   ..
                                  (Signature of official issuing the
                                  Certificate)

(Seal or stamp of the Issuing Authority, as appropriate)

Under the provisions of Regulation 7(2) and (4) of Annex IV of the Convention the
validity of this Certificate is extended until
....

                          Signed ....
                          (Signature of duly authorized official)
                          Place ....
                          Date ....

(Seal or stamp of the Authority, as appropriate)

---

* Delete as appropriate
** Parameters should be incorporated

ANNEX V

REGULATIONS FOR THE PREVENTION OF POLLUTION BY GARBAGE FROM SHIPS

Regulation 1

Definitions

For the purposes of this Annex:

(1) "Garbage" means all kinds of victual, domestic and operational waste excluding fresh fish and parts thereof, generated during the normal operation of the ship and liable to be disposed of continuously or periodically except those substances which are defined or listed in other Annexes to the present Convention.

(2) "Nearest land". The term "from the nearest land" means from the baseline from which the territorial sea of the territory in question is established in accordance with international law except that, for the purposes of the present Convention "from the nearest land" off the north eastern coast of Australia shall mean from a line drawn from a point on the coast of Australia in

latitude $11°00'$ South, longitude $142°08'$ East to a point in
 latitude $10°35'$ South,
longitude $141°55'$ East, thence to a point latitude $10°00'$ South,
longitude $142°00'$ East, thence to a point latitude $9°10'$ South,
longitude $143°52'$ East, thence to a point latitude $9°00'$ South,
longitude $144°30'$ East, thence to a point latitude $13°00'$ South,
longitude $144°00'$ East, thence to a point latitude $15°00'$ South,
longitude $146°00'$ East, thence to a point latitude $18°00'$ South,
longitude $147°00'$ East, thence to a point latitude $21°00'$ South,
longitude $153°00'$ East, thence to a point on the coast of Australia
 in latitude $24°42'$ South, longitude $153°15'$ East.

(3) "Special area" means a sea area where for recognized technical reasons in relation to its oceanographical and ecological condition and to the particular character of its traffic the adoption of special mandatory methods for the prevention of sea pollution by garbage is required. Special areas shall include those listed in Regulation 5 of this Annex.

Regulation 2

Application

The provisions of this Annex shall apply to all ships.

Regulation 3

Disposal of Garbage outside Special Areas

(1) Subject to the provisions of Regulations 4, 5 and 6 of this Annex:

 (a) the disposal into the sea of all plastics, including but not limited to synthetic ropes, synthetic fishing nets and plastic garbage bags is prohibited;

 (b) the disposal into the sea of the following garbage shall be made as far as practicable from the nearest land but in any case is prohibited if the distance from the nearest land is less than:

  (i) 25 nautical miles for dunnage, lining and packing materials which will float;

  (ii) 12 nautical miles for food wastes and all other garbage including paper products, rags, glass, metal, bottles, crockery and similar refuse;

 (c) disposal into the sea of garbage specified in sub-paragraph (b)(ii) of this Regulation may be permitted when it has passed through a comminuter or grinder and made as far as practicable from the nearest land but in any case is prohibited if the distance from the nearest land is less than 3 nautical miles. Such comminuted or ground garbage shall be capable of passing through a screen with openings no greater than 25 millimetres.

(2) When the garbage is mixed with other discharges having different disposal or discharge requirements the more stringent requirements shall apply.

Regulation 4

Special Requirements for Disposal of Garbage

(1) Subject to the provisions of paragraph (2) of this Regulation, the disposal of any materials regulated by this Annex is prohibited from fixed or floating platforms engaged in the exploration, exploitation and associated offshore processing of sea-bed mineral resources, and from all other ships when alongside or within 500 metres of such platforms.

(2) The disposal into the sea of food wastes may be permitted when they have been passed through a comminuter or grinder from such fixed or floating platforms located more than 12 nautical miles from land and all other ships when alongside or within 500 metres of such platforms. Such comminuted or ground food wastes shall be capable of passing through a screen with openings no greater than 25 millimetres.

Regulation 5

Disposal of Garbage within Special Areas

(1) For the purposes of this Annex the special areas are the Mediterranean Sea area, the Baltic Sea area, the Black Sea area, the Red Sea area and the "Gulfs area" which are defined as follows:

(a) The Mediterranean Sea area means the Mediterranean Sea proper including the gulfs and seas therein with the boundary between the Mediterranean and the Black Sea constituted by the 41°N parallel and bounded to the west by the Straits of Gibraltar at the meridian of 5°36′W.

(b) The Baltic Sea area means the Baltic Sea proper with the Gulf of Bothnia and the Gulf of Finland and the entrance to the Baltic Sea bounded by the parallel of the Skaw in the Skagerrak at 57°44.8′N.

(c) The Black Sea area means the Black Sea proper with the boundary between the Mediterranean and the Black Sea constituted by the parallel 41°N.

(d) The Red Sea area means the Red Sea proper including the Gulfs of Suez and Aqaba bounded at the south by the rhumb line between Ras si Ane (12°8.5′N, 43°19.6′E) and Husn Murad (12°40.4′N, 43°30.2′E).

(e) The "Gulfs area" means the sea area located north west of the rhumb line between Ras al Hadd (22°30′N, 59°48′E) and Ras al Fasteh (25°04′N, 61°25′E).

(2) Subject to the provisions of Regulation 6 of this Annex:

(a) disposal into the sea of the following is prohibited:

(i) all plastics, including but not limited to synthetic ropes, synthetic fishing nets and plastic garbage bags; and

(ii) all other garbage, including paper products, rags, glass, metal, bottles, crockery, dunnage, lining and packing materials;

(b) disposal into the sea of food wastes shall be made as far as practicable from land, but in any case not less than 12 nautical miles from the nearest land.

(3) When the garbage is mixed with other discharges having different disposal or discharge requirements the more stringent requirements shall apply.

(4) Reception facilities within special areas:

(a) The Government of each Party to the Convention, the coastline of which borders a special area undertakes to ensure that as soon as possible in all ports within a special area, adequate reception facilities are provided in accordance with Regulation 7 of this Annex, taking into account the special needs of ships operating in these areas.

(b) The Government of each Party concerned shall notify the Organization of the measures taken pursuant to sub-paragraph (a) of this Regulation. Upon receipt of sufficient notifications the Organization shall establish a date from which the requirements of this Regulation in respect of the area in question shall take effect. The Organization shall notify all Parties of the date so established no less than twelve months in advance of that date.

(c) After the date so established, ships calling also at ports in these special areas where such facilities are not yet available, shall fully comply with the requirements of this Regulation.

Regulation 6

Exceptions

Regulations 3, 4 and 5 of this Annex shall not apply to:

(a) the disposal of garbage from a ship necessary for the purpose of securing the safety of a ship and those on board or saving life at sea; or

(b) the escape of garbage resulting from damage to a ship or its equipment provided all reasonable precautions have been taken before and after the occurrence of the damage, for the purpose of preventing or minimizing the escape; or

(c) the accidental loss of synthetic fishing nets or synthetic material incidental to the repair of such nets, provided that all reasonable precautions have been taken to prevent such loss.

Regulation 7

Reception Facilities

(1) The Government of each Party to the Convention undertakes to ensure the provision of facilities at ports and terminals for the reception of garbage, without causing undue delay to ships, and according to the needs of the ships using them.

(2) The Government of each Party shall notify the Organization for transmission to the Parties concerned of all cases where the facilities provided under this Regulation are alleged to be inadequate.

ATTACHMENT 3*

RESOLUTIONS ADOPTED BY THE INTERNATIONAL CONFERENCE ON MARINE POLLUTION, 1973

Resolution 1

Implementation of the 1969 Amendments to the International Convention for the Prevention of Pollution of the Sea by Oil, 1954

THE CONFERENCE,

NOTING its main objectives as set out in Resolution A.237(VII) adopted by the Assembly of the Inter-Governmental Maritime Consultative Organization on 12 October 1971, as being the achievement, by 1975 if possible but certainly by the end of the decade, of the complete elimination of the wilful and intentional pollution of the seas by oil and noxious substances other than oil and the minimization of accidental spills,

NOTING FURTHER Recommendation 86(e) of the United Nations Conference on the Human Environment, 1972, which called upon Governments to participate fully in the present Conference as well as in other efforts with a view to bringing all significant sources of pollution within the marine environment under appropriate controls, including in particular, the complete elimination of deliberate pollution by oil from ships with the goal of achieving this by the middle of the present decade,

RECOGNIZING the importance of the International Convention for the Prevention of Pollution of the Sea by Oil, 1954, as being the first multilateral instrument to be concluded with the prime objective of protecting the environment, and appreciating the significant contribution which that Convention has made in preserving the seas and coastal environment from pollution,

NOTING the Amendments to that Convention, set out in Resolution A.175(VI) adopted by the Assembly of the Organization on 21 October 1969, and considering that the implementation of those Amendments would be a major step towards the complete elimination of oil pollution and would bring about a significant reduction in the total quantity of oil reaching the sea,

BELIEVING that the International Convention for the Prevention of Pollution of the Sea from Ships, 1973, which was concluded by the present Conference will, when implemented, constitute a further important step towards the complete elimination of intentional pollution of the sea by harmful substances from ships,

BEING AWARE that some lapse of time will inevitably occur before the 1973 Convention can enter into force,

URGES Governments which have not yet accepted the 1969 Amendments to the International Convention for the Prevention of Pollution of the Sea by Oil, 1954, to do so as a matter of urgency without awaiting the entry into force of the International Convention for the Prevention of Pollution from Ships, 1973.

---

* For ATTACHMENT 2, see NOTE on page 3.

Resolution 2

Rapid Entry into Force of the International Convention for the Prevention of Pollution from Ships, 1973 and its Amendments

THE CONFERENCE,

BEING AWARE of the acuteness of the threat to the marine environment caused by pollution from ships,

HAVING DETERMINED to combat this form of pollution on the basis of and in accordance with the International Convention for the Prevention of Pollution from Ships, 1973, as adopted,

TAKING NOTE of paragraph (1) of Article 1 of this Convention by which the Parties to the Convention undertake to give effect to the provisions of the Convention and those Annexes thereto by which they are bound,

NOTES with particular interest Article 16 of the Convention which provides for a procedure accelerating the entry into force of amendments to Protocol I and to the Annexes and Appendices to the Convention,

REALIZES that the effectiveness of that amendment procedure largely depends on there being national procedures for rapid approval of amendments,

URGES States to become Parties to the Convention as soon as possible and to give effect to later amendments thereto with the minimum of delay.

Resolution 3

The Complete Elimination of Oil Pollution from Ships

THE CONFERENCE,

HAVING CONCLUDED the International Convention for the Prevention of Pollution from Ships, 1973,

BEING AWARE of Recommendation 86(e) adopted by the United Nations Conference on the Human Environment, 1972, recommending Governments, within the framework of the 1973 Inter-Governmental Maritime Consultative Organization Conference on Marine Pollution, *inter alia*, to strive towards complete elimination of deliberate pollution by oil from ships, with the goal of achieving this by the middle of the present decade,

NOTING that the Governing Council of the United Nations Environment Programme at its first session has requested the Executive Director to urge the Inter-Governmental Maritime Consultative Organization to set a time-limit for the complete prohibition of intentional oil discharge in the seas,

CONSIDERING that the Convention and particularly the regulations contained therein on the discharge of oil into the sea will be an important means of curbing pollution by oil from ships,

RECOGNIZING, however, that the Convention alone may not be sufficient for a satisfactory protection of the sea from pollution by oil from ships,

RECOMMENDS that Governments and other interested bodies concerned undertake concerted efforts, including the elaboration of additional regulations within the framework of the Organization and the provision of the necessary reception facilities, further to reduce the discharge of oil from ships into the sea with a view to the complete elimination of intentional pollution as soon as possible, but not later than the end of the present decade,

INVITES the Organization to take all possible measures to assist Governments in this task.

Resolution 4

Information on Penalties

THE CONFERENCE,

HAVING CONCLUDED the International Convention for the Prevention of Pollution from Ships, 1973,

NOTING that the penalties which shall be specified under the laws of the Parties to the Convention pursuant to Article 4 of this Convention must be adequate in severity to discourage violation of this Convention and must be equally severe irrespective of where the violation occurs,

CONSIDERING that each Party to this Convention has the sole competence to provide suitable penalties under its own laws,

RECOMMENDS that the Inter-Governmental Maritime Consultative Organization make available to all States Members of the Organization as well as Parties to the Convention information which might be relevant in considering a scale of suitable penalties applicable pursuant to Article 4 of the Convention.

Resolution 5

Intentional Pollution of the Sea and Accidental Spillages

THE CONFERENCE,

NOTING that it was assigned the following two objectives by Resolution A.237(VII), adopted by the Assembly of the Inter-Governmental Maritime Consultative Organization,

(1) the complete elimination of wilful and intentional pollution of the sea by oil and noxious substances other than oil; and

(2) the minimization of accidental spills;

these objectives to be achieved by 1975, if possible, but certainly by the end of the decade,

RECOGNIZING that it has primarily been as a result of extensive preparatory work within the Organization that the Conference has been able to prepare and open for signature the International Convention for the Prevention of Pollution from Ships, 1973,

BEING AWARE that the said Convention comprehensively covers the problem of intentional pollution by oil, noxious liquid substances in bulk, harmful substances in packaged forms or in freight containers or portable tanks or road and rail tank wagons, sewage and garbage, whereas it deals with the problem of accidental pollution only to a limited extent, bearing in mind that many aspects of this problem are and will continue to be dealt with within the framework of other technical Conventions relating to maritime safety,

BEING ALSO AWARE of the close relationship between ship safety and the prevention of pollution from ships,

RECOGNIZING ALSO that considerable progress has been made by the Organization in furtherance of the second objective, by developing proposed international rules and standards directed towards, or contributing to, the prevention, mitigation and minimization of accidental pollution, including the prevention of accidents to ships, minimization of spillages after accident and mitigation of damage after spillages,

RECOGNIZING FURTHER that a considerable amount of work in this field leading to the formulation of, and amendments to, conventions for which the Organization is depositary, and other instruments relating to ship safety and prevention of pollution, has yet to be accomplished,

RECOMMENDS that the Organization pursue and encourage studies relating to pollution abatement in the marine environment such as:

(a) collection of scientific data on the identification of harmful substances transported by ships and their effect on the marine environment;

(b) collection of ship casualty statistics, particularly on casualties resulting in the pollution of the marine environment;

(c) analysis of such casualty data including the interrelationship of average tanker size and age with incidents and magnitude of pollution casualties,

RECOMMENDS FURTHER that the Organization continue its work with high priority on the development of measures for the minimization of accidental spillages, particularly those relating to:

(a) prevention of accidents to ships including:

(i) safe navigational procedures and traffic separation schemes for the prevention of collisions, strandings and groundings, this to include the ultimate development of international performance standards for navigational aids;

(ii) watchkeeping practices in port and at sea and the training and certification of seamen;

(iii) provision of modern navigational and communications equipment;

(iv) the operational procedures during the transfer, loading and unloading of oil and noxious substances;

(v) manoeuvrability and controllability of large ships;

(vi) construction and equipment of ships carrying oil or noxious substances; and

(vii) safe carriage of dangerous goods in packaged forms or in freight containers or portable tanks or road and rail tank wagons,

(b) minimization of the risk of escape of oil and other noxious substances in the event of maritime accidents, including facilitation of transfer of cargo in the event of accidents,

(c) minimization of pollution damage to the marine environment including:

(i) study and development of new techniques and methods for cleaning, recycling and disposing of hazardous substances carried by ships; and

(ii) technical study and development of devices and chemicals used in removing oil and other harmful substances discharged into the sea,

with a view to having appropriate action taken by way of the adoption and implementation at an early date of amendments to existing conventions relating to safety at sea and prevention of pollution or of new conventions, as appropriate.

Resolution 6

Control of Discharge of Oil

THE CONFERENCE,

NOTING that all petroleum-derived oils are regulated under Annex I of the International Convention for the Prevention of Pollution from Ships, 1973,

NOTING FURTHER that the regulation of certain light refined petroleum oils under Annex I of the Convention introduces a new dimension and scope to international control of ship-generated oil pollution,

RECOGNIZING that different types of petroleum-derived oils may behave differently in the marine environment and may have different hazard characteristics, and

CONSIDERING that the behaviour and effects of all petroleum-derived oils in the marine environment, and in particular the methods and procedures for controlling their discharge from ships, are appropriate matters for further study by the Inter-Governmental Maritime Consultative Organization,

RECOMMENDS that the Organization take appropriate steps, at an early date, to review, on a comprehensive basis, the environmental problems created by the discharge of all petroleum-derived oils into the marine environment, with particular reference to the problems associated with the discharge of light refined oils and with a view to possible improvement of the provisions of Annex I of the Convention.

Resolution 7

Method to Identify the Source of Discharged Oil

THE CONFERENCE,

HAVING IN MIND Regulation 9 of Annex I of the International Convention for the Prevention of Pollution from Ships, 1973, whereby the discharge of oil or oily mixtures from ships shall be prohibited except when such discharge satisfies specified conditions,

RECOGNIZING the need to ensure that any ship which has discharged oil or oily mixtures in contravention of the said Regulation shall be identified promptly and punished,

RECOGNIZING ALSO that some Governments have promoted work to develop a practical method whereby the discharged oil can be promptly identified as the oil loaded on board a certain ship,

URGES those Governments to continue their efforts and all other Governments to initiate research into this problem, with a view to arriving at an early solution.

Resolution 8

Draught Requirements for Segregated Ballast Tankers

THE CONFERENCE,

NOTING that Regulation 13 of Annex I of the International Convention for the Prevention of Pollution from Ships, 1973, in determining the amount of required segregated ballast capacity, specifies a segregated ballast draught as a function of ship length, and that this will be applied to tankers of 150 metres in length and above,

NOTING FURTHER that this requirement is largely based on experience which pertains in general to large tankers where the amount of ballast taken aboard has been left to the discretion of the Master,

RECOMMENDS that the Inter-Governmental Maritime Consultative Organization take appropriate action to consider these ballast draught requirements, taking full account of further experience with ships of various sizes which have operated safely in their ballast conditions and to examine them with a view to determining whether any improvement is required, with special regard to the need for a more specific requirement for tankers of less than 150 metres in length.

Resolution 9

Tonnage Measurement of Segregated Ballast Oil Tankers

THE CONFERENCE,

NOTING that Regulation 13 of Annex I of the International Convention for the Prevention of Pollution from Ships, 1973 requires segregated ballast for new oil tankers of 70,000 tons deadweight and above,

NOTING ALSO that this requirement may cause new segregated ballast oil tankers to have substantial increases in freeboard and certain principal dimensions, in comparison with existing oil tankers, for equivalent productive cargo deadweights,

NOTING FURTHER that substantially increased principal dimensions without increased deadweight may in some cases increase either gross or net registered tonnage or both, for segregated ballast oil tankers,

RECOMMENDS that the Inter-Governmental Maritime Consultative Organization study the matter of equitable determination of gross and net registered tonnage for segregated ballast oil tankers in comparison with existing oil tankers of equivalent productive cargo deadweight.

Resolution 10

Development of Efficient Oil Content Monitoring Arrangements

THE CONFERENCE,

NOTING that the Regulations contained in Annex I of the International Convention for the Prevention of Pollution from Ships, 1973 rely for their control and enforcement in a number of instances on an oil discharge monitoring system and, in particular, that Regulation 15 of that Annex requires that an oil tanker designed for retention of oil on board shall be fitted with such a system to control the quality of any effluent discharged into the sea,

NOTING ALSO that Regulation 1(16) of that Annex provides for ballast to be considered as clean ballast if oil content monitoring arrangements establish that the oil content of the effluent from such a tank does not exceed 15 parts per million,

NOTING FURTHER the Recommendation on International Performance Specifications for Oily-Water Separating Equipment and Oil Content Meters adopted by the Assembly of the Inter-Governmental Maritime Consultative Organization by Resolution A.233(VII),

RECOGNIZING that further progress in the development of such monitors is an urgent requirement,

RECOMMENDS that the Organization should promote studies with a view to developing more sensitive, accurate and reliable oil content measuring instruments to cope with the full range of the oils covered by that Annex.

Resolution 11

Limitation of Size and Arrangement of Cargo Tanks in Oil Tankers

THE CONFERENCE,

NOTING with satisfaction that most tankers ordered since 1 January 1972 comply with the provisions regarding the limitation of the size and the arrangement of cargo tanks as laid down in the 1971 Amendments to the International Convention for the Prevention of Pollution of the Sea by Oil, 1954, contained in Resolution A.246(VII) adopted by the Assembly of the Inter-Governmental Maritime Consultative Organization, although those Amendments have not yet entered into force,

NOTING FURTHER that Resolution A.247(VII) of the Assembly of the Organization invites Governments to put into effect these requirements as soon as possible,

EMPHASIZING the desirability of the entry into force of the 1971 Amendments at the earliest possible date and in any case not later than the date of entry into force of the International Convention for the Prevention of Pollution from Ships, 1973,

BEING AWARE that some lapse of time will inevitably occur before the 1973 Convention can enter into force,

URGES all Governments to accept the Amendments to the 1954 Convention contained in Resolution A.246(VII) of the Assembly of the Organization as soon as possible.

Resolution 12

Development of Scientific Information on Water Quality Criteria

THE CONFERENCE,

RECOGNIZING that the capacity of the sea to assimilate pollutants and render them harmless is limited and that its ability to regenerate natural resources is also limited,

BELIEVING that the adequacy of measures taken to prevent pollution of the sea by substances that are liable to create hazard to human health, to harm marine life, to damage amenities or to interfere with other legitimate uses of the sea needs to be kept under review,

BELIEVING ALSO that there is a need to organize all interested competent organizations in establishing methods whereby the needs of the marine environment relative to water quality can be established, to identify the sources of pollution and continually assess the various methods of controlling marine pollution for the development of new or more effective control measures where appropriate,

RECOMMENDS that the Inter-Governmental Maritime Consultative Organization should co-operate with other organizations and in particular with the Joint Group of Experts on the Scientific Aspects of Marine Pollution (GESAMP) to achieve these aims whereby a first step might be to examine the method and procedure necessary to establish water quality criteria for the protection of the marine environment.

Resolution 13

Procedures and Arrangements for the Discharge of
Noxious Liquid Substances into the Sea

THE CONFERENCE,

HAVING CONCLUDED, in pursuance of its main objectives, the International Convention for the Prevention of Pollution from Ships, 1973, which, *inter alia*, contains in Annex II Regulations for the Control of Pollution by Noxious Liquid Substances in Bulk,

NOTING, in particular, Regulation 5 of Annex II by which the discharge into the sea of noxious liquid substances of Categories A, B, C and D or of ballast water, tank washings or other residues or mixtures containing such substances is prohibited, except in compliance with specified conditions including procedures and arrangements which shall be approved by the Administration to ensure that the criteria specified for each Category will be met,

DESIRING to facilitate international trade by ensuring, as far as possible, the uniform implementation of Annex II,

RECOMMENDS that the Inter-Governmental Maritime Consultative Organization should ensure, with a view to providing a uniform basis for the guidance of the Parties to the Convention in approving such procedures and arrangements, that the necessary studies are undertaken with highest priority, in order to develop the standards referred to in Regulations 5 and 8 of Annex II,

RECOMMENDS FURTHER that the Organization should subsequently review the form of the Cargo Record Book contained in Appendix IV of Annex II of the Convention, taking into account the standards for procedures and arrangements previously developed.

Resolution 14

Recommendation on Hazard Evaluation of Noxious Liquid Substances

THE CONFERENCE,

HAVING CONCLUDED, in pursuance of its main objectives, the International Convention for the Prevention of Pollution from Ships, 1973, which, *inter alia*, contains in Annex II Regulations for the Control of Pollution by Noxious Liquid Substances in Bulk,

NOTING Resolution 17 by which the Conference recommended the development of appropriate provisions relating to the control of pollution by noxious solid substances carried in bulk,

NOTING, in particular, Regulations 3 and 4 of Annex II and its Appendices II and III by which liquid substances are categorized in accordance with their environmental hazards when released into the sea through the normal operation of ships,

NOTING ALSO with appreciation that the Joint Group of Experts on the Scientific Aspects of Marine Pollution (GESAMP) had developed a rationale and had made hazard evaluations of some 400 substances which provided a sound scientific basis for their categorization,

DESIRING to facilitate international trade by avoiding, as far as possible, the necessity for Parties to the Convention to enter into consultation on substances not listed in Appendices II and III to Annex II,

NOTING FURTHER, however, that there are substances which require further data in order to complete the evaluation of their environmental hazards, particularly in relation to living resources,

BEING AWARE of the need to keep these Appendices up to date,

RECOMMENDS that the Inter-Governmental Maritime Consultative Organization should as a matter of urgency take appropriate steps:

(a)   to review the criteria used to define Category D substances;

(b) to evaluate the hazard of those substances for which further data were found necessary as well as new substances proposed to be carried in accordance with the rationale developed by GESAMP; and

(c) to increase all the lists to cover all the substances known to be carried,

INVITES Governments to pursue and encourage studies on environmental hazards of such substances and provide the Organization, as specified in the Annex to this Resolution, with as much information as is available.

ANNEX TO RESOLUTION 14

Information on a New Substance to be Transported by Ships for the Evaluation of its Environmental Hazards

1. Correct technical name: ....

   (Secondary or alternative name(s)) ....

Note: The information listed below would enable a complete assessment to be made but a provisional assessment may be based on as much relevant information as is currently available to the Governments involved.

2. Chemical formula: ....

3. Physical properties:

   (a) Boiling point: ....°C

   (b) Melting point: ....°C

   (c) Specific gravity: ....

   (d) Vapour pressure: .... kp/cm² at 37.8°C

   (e) Solubility in water: .... mg/1 at 20°C

   (f) Viscosity:.....................................

   (g) Odour (qualitative description): ....

   (h) Colour: ....

4. Chemical and biochemical properties:

   (a) Chemical stability (oxidation, reduction, UV light): ....

   (b) Reactivity with sea water:.....................................
   ....

   (c) Biodegradability:..

   (d) Chemical oxygen demand (COD)/5-day Biochemical oxygen demand (BOD)5 .... mg/1 (20°C)

   (e) Biotransformation (where known): ....

   (f) Polymerizability under exposure to the atmosphere and sunlight: ....

   (g) Lipid solubility: ....

5. Bioaccumulation by marine organisms (cf. GESAMP IV/19/Supp.1, paragraphs 23-26):

   (a) Rate and level of uptake and retention of substances: ....

   (b) Tainting effect: ....

   (c) Colour and other appearance changes: ....

6. Other damage to marine living resources (cf. GESAMP IV/19/Supp.1, paragraphs 27-30) Toxicity (TLm$_{96}$): .... ppm.
   ....

7. Hazard to human health (cf. GESAMP IV/19/Supp.1, paragraphs 31-34, 37):

   (a) By oral intake: .... mg/kg (LD$_{50}$)

   (b) By skin contact and inhalation: ....

8. Effect on amenities (cf. GESAMP IV/19/Supp.1, paragraphs 38-42):
   ....

9. Additional remarks (briefly describe test conditions for items 5, 6 and 7 above).

Note: Approved standard method should be used where possible.

Resolution 15

Recommendation Concerning the Convention Provisions Relating to the Carriage of Noxious Liquid Substances in Bulk

THE CONFERENCE,

NOTING the Regulations relating to the design, construction, equipment and procedures for ships carrying noxious liquid substances in bulk contained in Annex II of the International Convention for the Prevention of Pollution from Ships, 1973, in particular Regulation 13(2) of that Annex by which Parties to the Convention are obliged to issue, or to cause to be issued, detailed requirements on the design, construction, equipment and procedures for such ships in order to ensure compliance with Regulation 2(1) of that Annex,

NOTING ALSO Regulation 13(3) of that Annex which requires that for chemical tankers the detailed requirements shall contain at least all the provisions given in the Code for the Construction and Equipment of Ships Carrying Dangerous Chemicals in Bulk ("Bulk Chemical Code") adopted by the Assembly of the Inter-Governmental Maritime Consultative Organization in Resolution A.212(VII),

NOTING FURTHER that the Organization has prepared an approach to modification of the Bulk Chemical Code to include marine pollution prevention measures,

DESIRING the formulation of appropriate provisions for the carriage of noxious liquid substances in bulk in ships that are not self-propelled and in ships other than chemical tankers,

RECOMMENDS that the Organization:

(a) amends the Bulk Chemical Code as early as possible in order to include requirements necessary from the marine pollution prevention point of view and also to ensure consistency with the provisions of the Convention, in particular the definition of a new and existing ship in paragraph 1.7 of the Code;

(b) keeps the Code under constant review with regard to prevention of marine pollution, taking into account both experience and future development of technology; and

(c) develops with priority Codes for the carriage of noxious liquid substances in bulk in ships that are not self-propelled and in ships other than chemical tankers.

Resolution 16

Recommendation Concerning the Prevention of Pollution
by Liquefied or Compressed Gases Carried in Bulk

THE CONFERENCE,

NOTING that the International Convention for the Prevention of Pollution from Ships, 1973, contains in Annex II Regulations for the Control of Pollution by Noxious Liquid Substances in Bulk which are framed to eliminate or minimize intentional or accidental pollution by such substances,

RECOGNIZING a potential hazard to the environment in general which is also involved in the carriage of some liquefied or compressed gases in bulk by ships,

NOTING ALSO that the Inter-Governmental Maritime Consultative Organization has under preparation a Code for the Construction and Equipment of Ships Carrying Dangerous Liquefied or Compressed Gases in Bulk ("Gas Carrier Code"),

RECOMMENDS that:

(a) the Organization should use all its endeavours to bring the Gas Carrier Code to the earliest possible completion; and

(b) Parties to the Convention, following the finalization of the Gas Carrier Code, should issue or cause to be issued such national requirements as may be necessary to minimize any harmful effect of transporting liquefied or compressed gases in bulk on the environment.

Resolution 17

Recommendation Concerning the Prevention of Pollution
by Noxious Solid Substances Carried in Bulk

THE CONFERENCE,

NOTING that the International Convention for the Prevention of Pollution from Ships, 1973, contains in Annex II Regulations for the Control of Pollution by Noxious Liquid Substances in Bulk which are framed to eliminate or minimize intentional or accidental pollution by such substances,

RECOGNIZING a potential hazard to the marine environment which is also involved in the carriage of noxious solid substances in bulk by ships,

RECOGNIZING ALSO a possible need to formulate appropriate provisions for inclusion in the International Convention for the Prevention of Pollution from Ships, 1973,

NOTING however that the present state of knowledge in this field has not advanced sufficiently to enable the Conference to formulate such provisions,

RECOMMENDS that:

(a) the Inter-Governmental Maritime Consultative Organization pursue and encourage studies on the impact that the carriage of noxious solid substances in bulk by ships may have upon the marine environment and on the measures for minimizing the threat to the marine environment which arises from the carriage of such substances; and

(b) the results of such studies be directed towards the development of the appropriate provisions relating to the control of pollution by noxious solid substances carried in bulk for inclusion in the 1973 Convention.

INVITES Governments:

(a) to forward reports of incidents involving noxious solid substances carried in bulk by ships to the Organization pending development of the regulations of the 1973 Convention; and

(b) to issue, or cause to be issued, such national requirements as may be necessary to minimize any harmful effect of transporting noxious solid substances in bulk on the environment.

Resolution 18

Research into the Effect of Discharge of Ballast Water
Containing Bacteria of Epidemic Diseases

THE CONFERENCE,

NOTING that ballast water taken in waters which may contain bacteria of epidemic diseases may, when discharged into the sea in another location, cause a danger of spreading of the epidemic diseases to other countries,

REQUESTS the World Health Organization, in collaboration with the Inter-Governmental Maritime Consultative Organization, to initiate studies on that problem on the basis of any evidence and of proposals which may be submitted by any Government.

Resolution 19

Recommendation Concerning the Prevention of Pollution by Harmful Substances Carried by Sea in Packaged Forms or in Freight Containers, Portable Tanks or Road and Rail Tank Wagons

THE CONFERENCE,

NOTING the Regulations set forth in Annex III of the International Convention for the Prevention of Pollution from Ships, 1973, relating to the carriage of harmful substances by sea in packaged forms, or in freight containers, portable tanks, or road and rail tank wagons, in particular Regulation 1(3) of that Annex by which Parties to the Convention are obliged to issue, or cause to be issued, detailed instructions on packaging, marking and labelling, documentation, stowage, quantity limitations, exceptions and notification for preventing or minimizing pollution of the marine environment,

NOTING ALSO the Regulations relating to the safe carriage of dangerous goods by sea as set out in Chapter VII of the International Convention for the Safety of Life at Sea, 1960, in particular Regulation 1(d) of that Chapter by which Contracting Governments are obliged to issue, or cause to be issued, detailed instructions for the safe packing and stowage of specific dangerous goods or categories of dangerous goods which shall include any precautions necessary in relation to other cargo,

NOTING FURTHER the International Maritime Dangerous Goods Code which was prepared in implementation of Recommendation 56 of the International Conference on Safety of Life at Sea, 1960, and has been recommended by the Inter-Governmental Maritime Consultative Organization as a uniform basis upon which Governments should formulate the national regulations envisaged in Chapter VII of the 1960 Safety Convention,

RECOGNIZING that provisions concerning harmful substances as defined in Article 2(2) of the 1973 Convention must be specified and be complementary to those which have been adopted for the carriage of dangerous goods by sea,

RECOMMENDS that:

(a) the Organization pursue and encourage studies on the impact that the carriage by sea of such harmful substances in packaged forms, or in freight containers, portable tanks, or road and rail tank wagons, may have upon the marine environment;

(b) the results of such studies be directed towards the revision of the scope of the International Maritime Dangerous Goods Code, taking into account:

   (i) substances that are harmful to the marine environment whether or not they are classed as dangerous goods;

   (ii) the minimization of the threat to the marine environment that arises from the carriage by sea of the substances that will be enumerated in that Code; and

   (iii) safety in maritime transport;

(c) in such revision particular account be taken of:

   (i) packaging,

   (ii) marking and labelling,

   (iii) documentation,

   (iv) stowage,

   (v) quantity limitations,

   (vi) exceptions, and

   (vii) notification.

(d) Governments consider adoption of the format of the International Maritime Dangerous Goods Code for the systematic development of regulations and standards for the carriage by sea of harmful substances that represent a threat to the marine environment so as to ensure compatibility between safety requirements and provisions relating to pollution abatement;

(e) such particulars as referred to in this paragraph form the basis for the further development of the provisions of the Regulations contained in Annex III of the 1973 Convention; and

(f) Parties to the 1973 Convention make arrangements to cater for the possible need to recover or otherwise deal with harmful substances which are lost or may be lost into the sea from ships.

Resolution 20

Provision of Standards and Test Methods
Concerning Discharge of Sewage

THE CONFERENCE,

NOTING that Annex IV of the International Convention for the Prevention of Pollution from Ships, 1973, contains certain requirements concerning the discharge of sewage into the sea from ships which should be based on standards and test methods to be developed by the Inter-Governmental Maritime Consultative Organization,

URGES the Organization to take action to develop such standards and test methods as soon as possible.

Resolution 21

Provision of Reception Facilities for the Discharge of Sewage and Disposal of Garbage

THE CONFERENCE,

NOTING that Annexes IV and V of the International Convention for the Prevention of Pollution from Ships, 1973, provide that the discharge of sewage and disposal of garbage into the sea from ships shall be prohibited except when specified conditions are satisfied,

RECOGNIZING the need for adequate reception facilities to make possible the application of these requirements for the discharge of sewage and disposal of garbage,

RECOGNIZING FURTHER that the effective implementation of Annexes IV and V of the Convention is dependent upon the availability of such reception facilities on a world-wide basis,

URGES Governments to take appropriate action to ensure the provision, as early as possible, of adequate facilities for the reception of sewage and garbage from ships, adequate to meet the needs of the ships using them without causing undue delay.

Resolution 22

Promotion of Technical Co-operation

THE CONFERENCE,

RECOGNIZING that the complete elimination of pollution in the marine environment by ships requires broad international co-operation and technical and scientific resources,

RECOGNIZING FURTHER that Parties to the International Convention for the Prevention of Pollution from Ships, 1973, will be asked to undertake full responsibility and make arrangements for detecting, monitoring and preventing or mitigating pollution by ships,

BELIEVING that the promotion of technical co-operation on an intergovernmental level will hasten the implementation of the Convention by States not already possessing the necessary or adequate technical and scientific expertise,

URGES Governments to promote, in consultation with the Inter-Governmental Maritime Consultative Organization and other international bodies, and with assistance and co-ordination by the Executive Director of the United Nations Environment Programme, support for those States which request technical assistance for:

(a) the training of scientific and technical personnel;

(b) the supply of necessary equipment and facilities for monitoring;

(c) the facilitation of other measures and arrangements to prevent or mitigate pollution of the marine environment by ships; and

(d) the encouragement of research,

URGES FURTHER Governments to initiate action in connexion with the above without awaiting the entry into force of the Convention.

Resolution 23

Nature and Extent of States' Rights over the Sea

THE CONFERENCE,

BEARING IN MIND that a United Nations Conference on the Law of the Sea is to be convened pursuant to Resolution 2750 C (XXV) of the General Assembly of the United Nations,

TAKING INTO ACCOUNT the specialized character of the present Conference,

CONSIDERING that the International Convention for the Prevention of Pollution from Ships, 1973, establishes technical requirements relating to the operation, design and equipment of ships with regard to the prevention of marine pollution, and that, wherever necessary, these international standards should be progressively amended and further improved within the framework of that Convention,

MINDFUL of paragraph (2) of Article 9 of the Convention,

NOTING that the Convention deals mainly with technical questions such as operation, equipment and design of ships,

BEING CONVINCED that the appropriate forum to deal with the question of the nature and extent of States' rights over the sea is the above-mentioned Conference on the Law of the Sea,

DECLARES that the decision of the present Conference reflects a clear intention to leave that question to the above-mentioned Conference on the Law of the Sea,

DECLARES FURTHER that the rights exercised by a State within its jurisdiction in accordance with the Convention do not preclude the existence of other rights of that State under international law.

Resolution 24

Co-ordination of Activities on the Prevention and Control of Marine Pollution

THE CONFERENCE,

NOTING that the International Convention for the Prevention of Pollution from Ships, 1973, has conferred upon the Inter-Governmental Maritime Consultative Organization and its Secretary-General, important functions to be performed under the Convention,

RECOGNIZING the need for effective co-ordination of activities carried out by different international organizations concerned with the prevention and control of marine pollution,

RECOMMENDS that the Organization, where necessary, consult with and seek assistance from other international organizations and expert bodies concerned within the United Nations system in order to achieve the objectives of the present Convention.

Resolution 25

Transmission of the International Convention for the Prevention of Pollution from Ships, 1973 to the United Nations Conference on the Law of the Sea

THE CONFERENCE,

BEARING IN MIND that a United Nations Conference on the Law of the Sea will be convened pursuant to Resolution 2750 C (XXV) of the General Assembly of the United Nations,

NOTING that, in accordance with the foregoing Resolution, international law concerning marine pollution forms a part of the Law of the Sea,

REQUESTS the Secretary-General of the Inter-Governmental Maritime Consultative Organization to forward the International Convention for the Prevention of Pollution from Ships, 1973, to the United Nations Conference on the Law of the Sea, so that this Convention can be taken into account in the broader context of that Conference.

Resolution 26

Establishment of the List of Substances annexed to the Protocol Relating to Intervention on the High Seas in Cases of Marine Pollution by Substances other than Oil

THE CONFERENCE,

NOTING that the Protocol relating to Intervention on the High Seas in Cases of Marine Pollution by Substances other than Oil, 1973, provides in its Articles I and III that the list of substances to be annexed to the Protocol shall be established and maintained by an appropriate body designated by the Inter-Governmental Maritime Consultative Organization,

NOTING FURTHER that the Protocol provides that Parties to the Protocol whether or not Members of the Organization shall be entitled to participate in the proceedings of the appropriate body when it considers matters relating to the list,

RECOGNIZING that the early establishment of this list will encourage acceptance of the Protocol by Governments and thereby promote the speedy entry into force of the Protocol,

REQUESTS the Organization to designate at the earliest practicable opportunity the appropriate body in accordance with the provisions of Articles I and III of the Protocol and to provide this body with the necessary facilities for its work,

REQUESTS that appropriate body to proceed with all speed and establish the list not later than 30 November 1974, which list shall be adopted by a two-thirds majority of those present and voting in that body,

RECOMMENDS that in establishing and maintaining the list of substances the appropriate body should consult and co-operate with competent international organizations,

REQUESTS the Secretary-General of the Organization, as soon as the list has been established, to annex copies thereof to the authentic texts of the Protocol,

REQUESTS FURTHER the Secretary-General of the Organization to communicate this list to Governments without delay.

LIST OF PERSONS ATTENDING THE CONFERENCE

The Government of the Argentine Republic

    Mr. R. M. Gowland
    Prefecto J. Durquet
    Dr. R. R. Lerena
    Mr. H. R. Basso
    Miss G. L. Grandi
    Mr. F. Mirré

The Government of the Commonwealth of Australia

    H.E. Mr. K. G. Brennan
    Mr. E. Willheim
    Mr. P. R. Holmes
    Mr. W. P. Crone
    Mr. T. B. Curtin
    Mr. P. G. Bassett
    Mr. W. B. Nicholson
    Dr. R. W. Greville

The Government of the State of Bahrain

    Dr. W. Al-Nimer

The Government of the Kingdom of Belgium

    Mr. R. Vancraeynest
    Mr. W. A. M. Bentein
    Mr. L. Van de Vel
    Mr. J. F. J. Van de Velde
    Mr. J. G. A. Gérard

The Government of the Federative Republic of Brazil

    Mr. Marcelo Raffaelli
    Mr. F. A. Fontoura
    Commander R. M. Tinoco
    Captain J. A. Munro
    Mr. C. M. Garcia
    Captain J. V. Fernandes
    Captain A. P. Torres
    Captain G. A. Barbosa
    Mr. P. R. M. Saboya
    Commander E. F. Vargas
    Mr. E. Nogueira de Sá

The Government of the People's Republic of Bulgaria

    H.E. Professor A. Yankov
    Mr. A. Valkanov
    Mr. G. Kandilarov
    Mr. P. Alexiev
    Mr. P. Kalthcev

The Government of the Byelorussian Soviet Socialist Republic

    Mr. V. Peshkov
    Mr. V. Borovikov

The Government of Canada

    The Honourable Jack Davis
    Mr. E. G. Lee
    Mr. R. R. MacGillivray
    Mr. T. C. Bacon
    Mr. R. W. Parsons
    Mr. L. H. Legault
    Dr. D. A. Munro
    Mr. D. L. Findlay
    Mr. J. F. Aspin
    Mr. T. G. Blyth
    Captain G. J. Davies
    Mr. A. Graham
    Mr. A. de Mestral
    Dr. R. W. Trites
    Dr. M. W. Waldichuck
    Dr. A. Walton
    Mr. B. M. Mawhinney

The Government of the Republic of Chile

    Vice-Admiral O. Buzeta
    Mr. E. Gómez
    Commander J. Sepúlveda

The Government of the Republic of Colombia

    Dr. A. Galeano

The Government of the Republic of Cuba

    Dr. J. Lopez García
    Captain S. C. Cancio
    Dr. M. Oporto Sánchez
    Dr. A. Bocourt Pérez
    Mr. O. A. Aguirre

The Government of the Republic of Cyprus

    Mr. A. D. Demetropoulos
    Mr. M. V. Vassiliades

The Government of the Kingdom of Denmark

    H.E. Mr. G. Seidenfaden
    Mr. T. Madsen
    Mr. L. Ravnebjerg
    Miss B. Poulsen
    Commander F. Otzen
    Mr. M. Bahn
    Mr. F. Lage
    Mr. H. V. Jacobsen
    Captain J. K. Højbjerg
    Captain H. Robdrup
    Mr. G. M. Falslev
    Mr. K. W. Linnemann
    Mr. B. Riise-Knudsen
    Mr. T. Holmstrøm
    Mr. J. J. Jensen

The Government of the Dominican Republic

    Mr. A. A. Ricart

The Government of the Republic of Ecuador

    Dr. J. Ortiz
    Commander F. Molina

The Government of the Arab Republic of Egypt

    Mr. M. A. El-Sammak
    Mr. S. Abdel-Hamid
    Dr. F. M. Ramadan
    Mr. M. Fawzi
    Mr. M. E. Abd-Rabbo

The Government of the Republic of Finland

    Mr. E. Helaniemi
    Mr. E. Yrjölä
    Mr. H. Muttilainen
    Mr. T. Niklander
    Mr. S. Hilden
    Mr. P. Tulkki
    Mrs. T. Melvasalo
    Mr. P. Forsskahl
    Mr. P. S. Gruner
    Mr. U. Turunen

The Government of the French Republic

    Mr. J. P. Cabouat
    Mr. J. Megret
    Mr. Leonard
    Mr. C. Douay
    Mr. J. Bigay
    Mr. Bothorel
    Miss J. Denoyel
    Mr. X. Duclaux
    Mr. P. Guerin
    Mr. M. Jacquier
    Mr. J. Lys
    Mr. F. G. A. Macart
    Mr. J.-C. Mourlon
    Mr. Nicolazo
    Mr. Pierani
    Mr. B. D. Roux
    Mrs. M. F. Watine

The Government of the German Democratic Republic

    Dr. H. Rentner
    Mr. A. Maul
    Mr. G. Mitschka
    Mr. R. Luther
    Dr. N. Trotz
    Mr. K. Winkel
    Professor H. Wünsche

The Government of the Federal Republic of Germany

    Dr. G. Breuer
    Mr. F. Stelter
    Mr. E. Happe
    Mr. A. Menzel
    Dr. A. Ritter von Wagner
    Dr. I. Joerss
    Mr. C. Boe
    Dr. J. Dorschel
    Mr. W. Häusler
    Mr. K.-J. Döhrn
    Mr. H. Hormann
    Mr. U. Bosse
    Mr. T. Becker
    Mr. M. Buck
    Mr. U. Ackermann
    Mr. W. Trennt
    Mr. K. Rolfs
    Mr. H.-J. Schröter

The Government of the Republic of Ghana

    H.E. Mr. H. V. H. Sekyi
    Mr. J. V. Gbeho
    Mrs. A. Y. Aggrey-Orleans
    Mr. E. S. Y. Dey
    Mr. M. D. Missinou
    Captain C. K. T. Dziworshie

The Government of the Republic of Greece

>Mr. N. Diamantopoulos
>Captain A. Chronopoulos
>Commander P. Kosmatos
>Lieutenant-Commander N. Kalyvas
>Mr. J. S. Perrakis
>Mr. G. Kolymvas
>Mr. J. Dimitracopoulos
>Mr. A. Tsiomis
>Mr. E. Verykokakis
>Mr. A. I. Chandris
>Mr. P. M. Nomikos
>Mr. Th. Nomikos
>Mr. Ch. Michalitsianos
>Mr. G. Timagenis
>Captain A. Batsis
>Mr. J. E. Kulukundis

The Government of the Republic of Haiti

>H.E. Dr. L. Mars
>Mr. M. Duplan

Hong Kong

>Mr. R. Blacklock
>Mr. F. C. Lingwood

The Government of the Hungarian People's Republic

>Mr. G. Kótai

The Government of the Republic of Iceland

>Mr. H. R. Bardarson
>Mr. H. Agustsson

The Government of the Republic of India

>Mr. S. V. Bhave
>Captain P. S. Vanchiswar
>Mr. D. A. Kamat
>Mr. S. Bannerjee
>Mr. I. J. Varma

The Government of the Republic of Indonesia

    Mr. M. Sjadzali
    Captain T. Surahardja
    Mr. R. U. Sukaton
    Mr. Nazif
    Mr. R. Sani
    Mr. Soempeno
    Mr. C. Tauran
    Mr. H. Saanin
    Mr. E. Soeprapto
    Mr. A. Soerja Djanegara

The Government of the Empire of Iran

    Mr. H. Afshar
    Dr. I. Said-Vaziri
    Mr. C. Vafaï
    Mr. A. Khajooei
    Mr. A. Farid
    Mr. A. Fathipour
    Mr. M. T. Farvar
    Mr. M. R. Amini
    Mr. M. Forootan

The Government of the Republic of Iraq

    H.E. Dr. A. Al-Kadhi
    Mr. A. Z. Ameen
    Mr. A. H. Jawad
    Mr. F. M. Al-Bayati

The Government of Ireland

    Miss M. C. Tinney
    Mr. T. Gorman
    Mr. P. D. Dempsey
    Mr. J. Flavin
    Mr. C. Lysaght
    Mr. D. Quigley
    Mr. J. Lynch
    Dr. T. McManus

The Government of the Italian Republic

    Mr. G. Pieraccini
    Ambassador C. Calenda
    Dr. A. Franchi
    Dr. L. Spinelli
    Professor E. Fasano
    Professor R. Passino
    Professor G. Albano
    Mr. F. Magi
    Dr. F. D'Aniello
    Mr. A. Luciano
    Mr. F. Bastianelli
    Dr. A. Basso
    Mr. U. Inzerillo
    Mr. B. Notari
    Mr. E. Rolle

The Government of the Republic of the Ivory Coast

    H.E. Mr. H. Polneau
    H.E. Mr. A. Traore
    Mr. B. Pegawagnaba
    Mr. G. K. Loukou

The Government of Jamaica

    Mr. D. J. Gayle

The Government of Japan

    H.E. Mr. S. Sugihara
    Mr. S. Harada
    Mr. T. Nakajima
    Mr. Y. Yoshinaga
    Mr. A. Kunibe
    Mr. M. Tada
    Mr. T. Mano
    Mr. K. Ichioka
    Mr. R. Tomura
    Mr. N. Saito
    Mr. M. Iwata
    Mr. H. Tani
    Mr. Y. Kawashima
    Mr. K. Iguchi
    Mr. M. Yoshida
    Mr. M. Ogushi
    Mr. R. Ishii

Mr. F. Yagawa
Mr. M. Koyama
Mr. T. Asahara
Mr. K. Yoshida
Mr. M. Abe
Mr. H. Fujii
Mr. H. Ibuki
Mr. M. Sanada
Mr. M. Hamada
Mr. H. Kobayashi
Mr. M. Watanabe
Mr. T. Taya
Mr. E. Tsuji
Mr. Y. Masuda
Mr. G. Yoshinaga
Mr. I. Usui

The Government of the Hashemite Kingdom of Jordan

Mr. A. G. Toukan
Mr. H. A. Nimah

The Government of the Republic of Kenya

Mr. R. O. Adero
Mr. F. Jackson
Mr. E. Dinga

The Government of the Khmer Republic

Mr. So Yandara

The Government of the State of Kuwait

H.E. Mr. A. A. Al-Naqib
Mr. A. R. M. Al-Yagout
Mr. A. S. Al-Mousa
Miss B. Al-Awadi
Mr. N. Al-Nakib

The Government of the Republic of Liberia

    The Honourable G. F. B. Cooper
    Mr. J. C. Montgomery
    Mr. W. E. Greaves
    Mr. H. N. Conway, Jr.
    Mr. F. T. Lininger
    Captain S. E. N. Pollock
    Dr. F. L. Wiswall, Jr.
    Mr. A. N. Narter
    Mr. D. B. Bannerman
    Mr. R. P. Harrison
    Mr. J. Wine
    Captain T. K. Yip
    Mr. J. M. Bates

The Government of the Libyan Arab Republic

    Mr. A. Y. El-Orfi

The Government of the Malagasy Republic

    Mr. M. Ramarozaka
    Mr. R. Rambahiniarison

The Government of the Republic of Malawi

    Mr. A. F. Mlota
    Mr. M. V. Phiri

The Government of the United Mexican States

    Dr. F. Vizcaíno Murray
    Dr. E. Echeverría Alvarez
    Mr. J. L. Vallarta
    Mr. J. Velasco
    Dr. E. Sánchez Palomera
    Commander G. López Lira
    Chief Engineer E. Urroz
    Chief Engineer C. Baptista
    Mr. F. Roux-López
    Mr. L. Bravo
    Dr. G. Martínez Narvaez
    Commander C. López Sotelo
    Chief Engineer M. A. García Lara
    Commander C. Maroto Gaxiola
    Miss S. Fuentes Berain
    Dr. F. Cesarman

The Government of the Principality of Monaco

    Mr. I. S. Ivanovic
    Professor J. Joseph
    Mr. A. Vatrican
    Mr. R. Projetti

The Government of the Kingdom of Morocco

    Mr. A. Majid

The Government of the Kingdom of the Netherlands

    Mr. J. F. van Doorn
    Mr. G. J. Boevé
    Mr. H. H. M. Sondaal
    Dr. I. A. Akrum
    Mr. C. van Dam
    Mr. D. Tromp
    Mr. J. A. Walkate
    Dr. J. H. Dewaide
    Mr. P. Bergmeijer
    Mr. H. F. M. Bertels
    Mr. E. H. Siccama
    Baron S. van Heemstra
    Mr. J. F. Lameijer
    Mr. L. A. S. Hageman
    Mr. B. Le Coultre

The Government of New Zealand

    Captain D. W. Boyes
    Mr. A. C. Doyle
    Mr. M. F. Watkins

The Government of the Federal Republic of Nigeria

    Mr. G. A. E. Longe
    Mr. O. A. Okanla
    Mr. O. Jemiyo
    Mr. A. O. Okafor
    Captain K. K. Laniyan
    Captain I. Y. O. Jonah
    Captain H. A. Agate

The Government of the Kingdom of Norway

    Mr. M. Hareide
    Mr. S. Storhaug
    Mr. E. Hareide
    Mr. H. Vindenes
    Mr. I. A. Manum
    Mr. P. Tresselt
    Mr. H. Chr. Bugge
    Mr. H. B. Hjelde
    Mr. J. E. Stang-Lund
    Mr. G. Stubberud
    Mr. Ø Frømyhr
    Mrs. A. V. Tjørswaag
    Mr. L. Føyn
    Mr. S. Øveraas
    Mr. J. W. Wilhelmsen, Jr.
    Mr. A. G. Offenberg
    Mr. B. H. Duborgh
    Mr. T. Riiser
    Mr. E. Archer
    Mr. A. Rikheim
    Mr. T. Honne
    Mr. T. Stoltenberg
    Mr. M. J. Thorp

The Government of the Sultanate of Oman

    H.E. Mr. N. S. El Bualy
    Dr. K. M. Hagras

The Government of the Republic of Panama

    Dr. R. A. Ehrman

The Government of the Republic of Peru

    H.E. Mr. A. Montagne
    Dr. J. Cacho-Sousa
    Commander L. H. Montes
    Mr. A. Rivero
    Lieutenant R. A. Forsyth

The Government of the Republic of the Philippines

    H.E. The Honourable J. Zobel de Ayala
    Mr. P. A. Araque
    Commodore E. R. Ogbinar
    Commissioner R. M. Lesaca
    Mr. M. C. Manansala
    Captain J. V. Francisco
    Mrs. S. Zaide Pritchard

The Government of the Polish People's Republic

    Mr. S. Perkowicz
    Mr. W. Ertel
    Mr. Z. Gandera
    Mr. P. Anders
    Mr. L. J. Tukasik
    Mr. W. Orszulok
    Mr. J. Siedlecki
    Mr. J. Jezioranski
    Mr. J. Potubinski
    Mr. W. Godlewski
    Mr. C. Dzienio

The Government of the Portuguese Republic

    Commander J. E. E. Cabido de Ataide
    Lieutenant-Commander M. J. D. C. Casquinho
    Mr. J. A. Belo
    Dr. C. Teixera da Motta

The Government of the Republic of Korea

    H.E. K. N. Choi
    Mr. J. I. Choi
    Mr. J. W. Roh
    Mr. K. B. Shin
    Mr. S. E. Kim
    Mr. H. Y. Chung
    Mr. C. K. Yoon
    Mr. J. H. Park
    Mr. W. K. Kim

The Government of the Republic of Viet-Nam

    Mr. V. Ninh

The Government of the Socialist Republic of Romania

    Captain V. Stan

The Government of the Kingdom of Saudi Arabia

    Mr. F. A. Basyoni
    Mr. A. Malki

The Government of the Republic of Singapore

    Mr. Koh Eng Tian
    Mr. Chee Kian Beng
    Mr. Lau Chee Peng

The Government of the Republic of South Africa

    Mr. J. J. Becker
    Mr. G. A. Visser
    Captain R. G. Gardner

The Government of the Spanish State

    H.E. Professor A. Poch
    Mr. A. Graiño
    Mr. R. Pastor
    Mr. J. Blanca
    Dr. A. Prego
    Mr. A. Oyarzábal
    Dr. J. A. de Yturriaga
    Mr. M. de la Hera
    Mr. A. Mato
    Dr. J. Ros
    Mr. P. Alvarez de Toledo
    Mr. E. Mirapeix

The Government of the Republic of Sri Lanka

    Mr. J. J. G. Amirthanayagam
    Mr. S C. A. Nanayakkara

The Government of the Kingdom of Sweden

   Mr. G. Steen
   Mr. G. Lind af Hageby
   Mr. L. Delin
   Mr. C. O. Senning
   Mr. G. Lindencrona
   Mr. P. Eriksson
   Mr. L. Danielson
   Mr. B. Looström
   Mr. J. B. Melchior
   Mr. G. Böös
   Captain L. Baecklund
   Mr. B. Stenström
   Mr. B. Dahlgren
   Mr. L. Kindahl
   Mr. S. Wiebe

The Government of the Swiss Confederation

   Mr. R. Bär
   Mr. R. Serex

The Government of the Kingdom of Thailand

   H.E. Mr. K. Suphamongkhon
   Group Captain W. Wiriyawit
   Mr. P. Charuchandr
   Senior Lieutenant P. Buranadilok

The Government of Trinidad and Tobago

   H.E. Dr. P. V. J. Solomon
   Mr. K. Ablack
   Mr. P. A. W. Hezekiah
   Mr. P. J. Dass

The Government of the Republic of Tunisia

   Mr. A. Turki
   Mr. S. Sedoucha
   Mr. B. Benzineb
   Mr. H. Boussoffara
   Mr. H. Cherif

The Government of the Republic of Turkey

    Mr. N. Dumlu

The Government of the Ukrainian Soviet Socialist Republic

    Mr. A. Tretiak
    Mr. E. Kachurenko
    Mr. A. Zinchenko

The Government of the Union of Soviet Socialist Republics

    Mr. V. Tikhonov
    Mr. V. S. Kotliar
    Mr. I. T. Matov
    Mr. P. N. Shternov
    Dr. G. P. Andrushaitis
    Mr. U. A. Antanaitis
    Dr. E. V. Borissov
    Dr. K. A. Velner
    Mr. V. P. Volokhov
    Mr. N. I. Glukhov
    Mr. B. F. Kasyanov
    Dr. A. L. Makovsky
    Dr. I. P. Miroshnitchenko
    Mr. S. M. Nunuparov
    Mr. N. N. Rodionov
    Mr. R. F. Sorokin
    Mr. V. V. Sutulo

The Government of the United Arab Emirates

    Dr. M. Mazzawi
    Mr. G. Al-Tajir

The Government of the United Kingdom of Great Britain and Northern Ireland

    Mr. J. N. Archer
    Mr. L. F. Standen
    Mr. G. Victory
    Mr. D. MacIver Robinson

Captain J. R. Hampton
Mr. C. Bell
Mr. N. Bell
Mrs. J. M. Wicks
Mr. E. W. G. Wilkins
Mr. W. H. Moore
Mr. A. F. Warner
Mr. R. K. Roberts
Mr. E. H. Whitaker
Dr. M. W. Holdgate
Mr. W. R. Small
Mr. H. A. Dudgeon
Mr. D. R. Morris
Mr. G. D. Crane
Mr. I. Prestt
Dr. D. L. Simms
Dr. E. J. Wilson
Mr. B. Hitch
Mr. D. H. Anderson
Mr. K. Chamberlain
Mr. M. R. Eaton
Mr. N. A. Smith
Mr. G. F. Buxton
Miss J. Grey
Mr. M. R. Phillips
Dr. H. A. Cole
Mr. A. Preston
Mr. P. C. Wood
Dr. J. E. Portmann
Dr. I. C. White
Dr. R. A. A. Blackman
Mr. R. W. Stewart
Mr. T. V. Cottam
Dr. P. G. Jeffery
Mr. H. J. McNeill
Mr. P. J. Sayers
Mr. C. W. M. Ingram
Mr. D. W. Lang
Miss V. J. Bell
Mr. P. R. Mitchell
Captain N. W. Finnis
Mr. L. Johnson
Mr. R. J. Pescod
Mr. C. W. Fyans
Mr. D. Butler
Mr. S. O. Ridgway
Mr. M. D. Squires
Mr. M. Elliston
Captain H. Blackmore
Mr. B. J. Brooke-Smith
Mr. R. J. C. Dobson
Mr. M. Anthony
Mr. O. G. Weller
Mr. F. H. Walmsley

Mr. W. G. Cann
Dr. L. Carter
Mr. H. Martin
Mr. J. Rigby
Mr. C. O. Jenkin-Jones
Mr. V. Sebek
Dr. G. Howells
Mr. A. R. Begg
Mr. D. Seaman
Mr. N. H. Baker
Mr. R. J. Furley
Mr. J. Wardley Smith

The Government of the United Republic of Tanzania

H.E. Mr. G. M. Nhigula
Mr. S. Ihema
Mr. J. L. Kateka

The Government of the United States of America

Mr. R. E. Train
Admiral C. R. Bender
Rear Admiral W. M. Benkert
Mr. B. H. Oxman
The Honorable J. Glenn Beall
The Honorable Warren G. Magnuson
The Honorable John M. Murphy
The Honorable Claiborne Pell
The Honorable Philip E. Ruppe
Captain H. H. Bell
Mr. T. C. Colwell
Mr. D. B. Cook
Mr. J. R. Enyart
Colonel F. Fedele
Mr. S. P. French
Mr. I. L. Fuller
Mr. R. K. Gregg
Mr. E. V. C. Greenberg
Captain L. W. Goddu, Jr.
Captain C. R. Hallberg
Mr. S. B. Jellinek
Mr. R. Johnson
Mr. R. J. Lakey
Mr. A. C. Landsburg
Mr. T. L. Leitzell
Mr. M. Matheson
Mr. R. McManus
Commander J. T. McQueston, Jr.
Mrs. C. Odell
Lieutenant Commander J. D. Porricelli

Mr. J. A. Reed
Mr. G. Ryan
Mr. H. P. Santiago
Captain K. B. Schumacher
Captain G. Steinman
Mr. H. A. Steyn, Jr.
Commander J. D. Sipes
Mr. H. D. Van Cleave
Captain S. A. Wallace
Lt. j.g. D. H. Williams
Captain P. A. Yost
Mr. J. Hussey
Mr. D. Keaney
Mr. C. L. Perian
Mrs. J. Perian
Mr. S. L. Sutcliffe
Mr. J. P. Walsh

The Government of the Eastern Republic of Uruguay

H.E. Mr. J. D. del Campo
Mr. A. Cazes

The Government of the Republic of Venezuela

H.E. Dr. C. Pérez de la Cova
Captain G. Nout
Mr. F. Marquez
Mr. R. Hernández

The Government of the Socialist Federal Republic of Yugoslavia

Dr. B. Sambrailo

United Nations

Mr. I. Steiner
Mr. G. Ivan Smith
Mr. F. Labastida

United Nations Environment Programme

Mr. M. F. Strong
Mr. K. Kaneko

Food and Agriculture Organization

    Mr. G. Moore

United Nations Educational, Scientific and Cultural Organization

    Dr. R. Griffiths

International Bank for Reconstruction and Development

    Mr. A. J. Carmichael

International Atomic Energy Agency

    Mr. M. Camcigil

European Economic Community

    Mr. S. Johnson
    Mr. H. Nagelmackers
    Mr. G. L. Close
    Miss M. Drabs
    Mr. R. Lugard-Brayne

International Institute for the Unification of Private Law

    Mr. M. Matteucci

International Chamber of Shipping

    Rear-Admiral P. W. W. Graham
    Mr. J. C. S. Horrocks
    Mr. J. M. S. Smith
    Mr. S. A. Cotton
    Mr. W. Welch
    Mr. N. R. McGilchrist
    Mr. S. K. Conacher
    Mr. M. P. Holdsworth
    Dr. A. L. Shrier
    Mr. R. C. Page
    Mr. B. Strenstrom
    Mr. I. E. Telfer
    Mr. W. N. M. Freeland
    Mr. D. N. Cleaver
    Commander T. Meyer

International Organization for Standardization

    Mr. C. Meredith
    Mr. A. A. B. Harvey

International Electrotechnical Commission

    Mr. G. Watson

International Union of Marine Insurance

    Mr. H. F. Duder

International Association of Ports and Harbors

    Commander D. Knight
    Captain G. Dudley
    Captain R. A. Gibbons
    Mr. A. J. Smith
    Mr. D. Drakley
    Mr. A. J. W. Harris

The Baltic and International Maritime Conference

    Mr. O. W. Arenfeldt
    Mr. P. G. F. Leader
    Miss M. Patten

International Association of Classification Societies

    Mr. F. H. Atkinson
    Mr. A. Shute
    Mr. R. Gardiner
    Mr. Engerrand

International Law Association

    Mr. R. B. Greenburgh
    Miss J. Gutteridge
    Dr. E. D. Brown

European Council of Chemical Manufacturers' Federations

    Mr. M. J. J. Bormans
    Mr. P. Schaeflé
    Mr. J. M. Watson
    Mr. Weber

Oil Companies International Marine Forum

    Mr. W. C. Brodhead
    Mr. C. A. Walder
    Captain I. E. LeCocq
    Mr. K. Nakayama
    Mr. J. R. Keates
    Mr. W. O. Gray
    Mr. R. J. Wheeler
    Mr. T. S. Wyman
    Mr. C. C. Smith

International Shipowners' Association

    Mr. I. Z. Bolioukh

Friends of the Earth International

    Mr. J. R. Sandbrook
    Mr. B. Johnson
    Miss A. Yurchyshyn

CONFERENCE SECRETARIAT

Secretary-General	Mr. Colin Goad
Deputy Secretary-General	Mr. J. Quéguiner
Executive Secretary	Captain A. Saveliev
Deputy Executive Secretary	Mr. Y. Sasamura
Deputy Executive Secretary	Dr. T. Mensah
Committee Secretary	Mr. T.S. Busha
Committee Secretary	Mr. S.L.D. Young
Committee Secretary	Mr. J. Jens
Committee Secretary	Mr. A. Andreev
Committee Secretary	Mr. B. Okamura
Assistant Committee Secretary	Mr. G. Cipolla
Assistant Committee Secretary	Mr. A. Spassky
Assistant Committee Secretary	Captain H. Wardelmann
Assistant Committee Secretary	Mr. F. de Franchis
Head of Conference Services	Mr. H. Mallet
Conference Officer	Miss R. Cadet
Documents Officer	Miss R. Heard
Press and Information Officer	Mrs. A. Meldrum